Build Your Own
Humanoid Robot

Build Your Own Humanoid Robot

Karl Williams

McGraw-Hill

New York   Chicago   San Francisco   Lisbon
London   Madrid   Mexico City   Milan   New Delhi
San Juan   Seoul   Singapore   Sydney   Toronto

The McGraw·Hill Companies

Cataloging-in-Publication Data is on file with the Library of Congress.

1 2 3 4 5 6 7 8 9 0   DOC/DOC   0 9 8 7 6 5 4

ISBN 0-07-142274-9

The sponsoring editor for this book was Judy Bass and the production supervisor was Pamela Pelton. It was set in Sabon by Patricia Wallenburg.

Printed and bound by RR Donnelley.

McGraw-Hill books are available at special quantity discounts to use as premiums and sales promotions, or for use in corporate training programs. For more information, please write to the Director of Special Sales, McGraw-Hill Professional, Two Penn Plaza, New York, NY 10121-2298. Or contact your local bookstore.

 This book is printed on recycled, acid-free paper containing a minimum of 50 percent recycled, de-inked fiber.

Contents

vi

Introduction

When I told my friends that I was writing a book about building humanoid robots, most of them commented that it would be nice to finally have a robot that could do some work around the house. They talked about a machine that could wash the dishes, take out the trash, vacuum the floor, prepare the meals, cut the grass, and tend the bar, etc. This is the kind of machine most people think of when they mention humanoid robots. My friends would also point out the amazing abilities of the Honda humanoid. Why aren't these machines available and working in our homes? I explained that it would cost millions of dollars and take a lot of time to build something comparable to the Honda humanoids.

Instead of building one complete humanoid robot, the projects in this book will cover some of the main components that make up such a machine. The fact is that the technologies necessary to build the projects in this book have become relatively inexpensive. With the right amount of imagination and innovation, anyone can create amazing machines in their basement laboratory. The robots of the future are now within our reach because we can build them ourselves! Robotics is a unique area of study because it encompasses many different disciplines such as electronics, computer science, mechanical design, control systems, programming, and biology. This is what makes building robots so interesting and fun. The humanoid robot fascinates humans because it is a machine that closely resembles life and humans themselves. I hope that this book inspires you to build the next generation of intelligent machines.

Acknowledgments

Thanks go to Judy Bass and the team at McGraw-Hill for all of their hard work, and to Patricia Wallenburg for doing a great job of putting the book together. Thanks go to my parents Gordon and Ruth Williams for all of their encouragement. To my brothers and their wives: Doug Williams, Gylian Williams, Geoff Williams, and Margaret Sullivan-Williams. Thanks go to the following people for their support and feedback: Stacey Dineen, Sachin Rao, Jennifer Reif, Laurie Borowski, John Lammers, Myke Predko, Jean Cockburn, Tom Cloutier, James Vanderleeuw, Darryl Archer, Paul Steinbach, Charles Cummins, Maria Cummins, Tracy Strike, Raymond Pau, Clark MacDonald, Rodi Snow, Sameer Siddiqi, Jack Kesselman, Steve Rankin, Randy Jones at Glitchbuster.com, and Bill Forgey at StampBuilder.com. Thanks also go to Jason Jackson, Roland Hofer, Kenn Booty, Patti Ramseyer, JoAnna Klueskens, and Tim Jones.

Tools, Test Equipment, and Materials

During the mechanical construction phase of building the projects in this book, a number of tools will be required. You will need a workbench or sturdy table in an area with good lighting. Try to keep your work area clean and free of clutter.

The first tool that will be used is the hacksaw. The hacksaw is designed to cut metal and hard plastics. When using the hacksaw to make straight cuts, it is a good idea to use a miter box. **Figure 1.1** shows the hacksaw (labeled L) and the miter box (K). If you have a little extra money and think you will be building a lot of robots, then you really need a band saw fitted with a metal cutting blade. The band saw shown in **Figure 1.2** is 9 inches, meaning that the saw can cut pieces up to a maximum length of 9 inches. This is perfect for building smaller robots, like the ones detailed in this book. With the metal cutting band saw, pieces of aluminum can be cut fast and with greater accuracy than a hacksaw.

An important piece of equipment that will be needed in your workshop is a vise, like the one shown in **Figure 1.3**. The vise will be needed quite often when cutting, drilling, and bending aluminum. Always clamp metal pieces tightly in the vise when working on them with other tools. It is dangerous to try drilling metal pieces that are not clamped in a vise.

You will need an electric drill during the mechanical construction phase of building the robot projects and the fabrication of the printed circuit boards. You will be required to drill approximately 150 holes during the process of creating each project in the book. An electric hand drill, like the one shown in **Figure 1.4**, can be used.

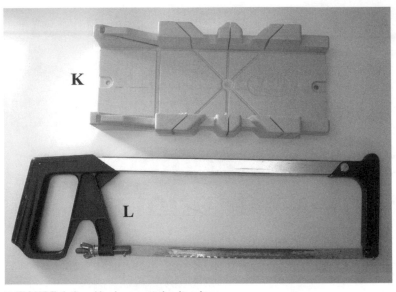

■ **FIGURE 1.1** *Hacksaw and miter box.*

■ **FIGURE 1.2** *Band saw fitted with a metal cutting blade.*

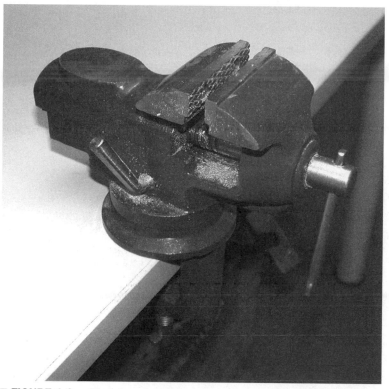

■ **FIGURE 1.3**   *Work bench vise.*

■ **FIGURE 1.4**   *Handheld electric drill.*

If you plan to build robots as a hobby, then purchasing a small drill press, like the one shown in **Figure 1.5**, would be a great idea. Using a drill press is highly recommended when drilling holes in printed circuit boards, where accuracy and straightness are important. These small drill presses don't cost much more than a good electric hand drill. I added an adjustable X-Y vise to the drill press in my workshop. This makes it possible to mill aluminum if an endmill, like the one shown in **Figure 1.6**, is purchased from a machine shop supplier. The drill press can then double as a small milling machine.

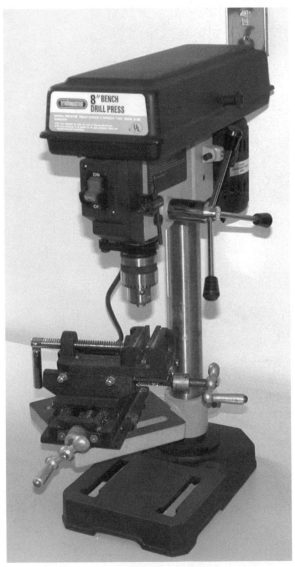

■ **FIGURE 1.5** *A small electric drill press with an X-Y adjustable vise.*

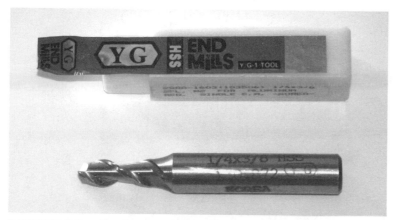

■ **FIGURE 1.6**   *Aluminum-cutting endmill.*

You will need a set of drill bits like the ones pictured in **Figure 1.7**. The 5/32-inch and 1/4-inch drill bits are used most often during the projects. You will need to separately buy the small 1/32-inch and 3/64-inch bits that will be used to drill the component holes in the printed circuit boards.

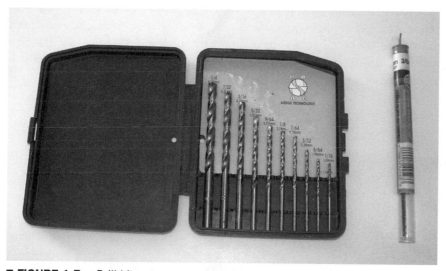

■ **FIGURE 1.7**   *Drill bit set.*

You will need an adjustable wrench (marked E in **Figure 1.8**), side cutters (F), pliers (G), needle nose pliers (H), a Phillips screwdriver (I), and a Robertson screwdriver (J) during construction of the robot projects. A set of miniature screwdrivers may be useful as well. The needle nose pliers can be used to hold wire and small components in place while soldering, bending wire, and holding machine screw nuts.

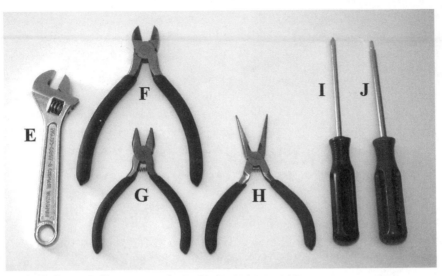

■ **FIGURE 1.8**   *Various pliers, a wrench, and screwdrivers.*

The wire strippers, shown in **Figure 1.9** (A), are used to strip the protective insulation off the wire, without cutting the wire itself. The device is designed to accommodate a number of wire sizes you will need. A pair of wire cutters (C) can cut wire when fabricating jumper wires and wiring power to the circuits. You will need rosin-core solder (B) when soldering components to the circuit boards, creating jumper wires, and wiring the battery connectors and power switches. To make soldering components to the printed circuit boards as easy as possible, buy the

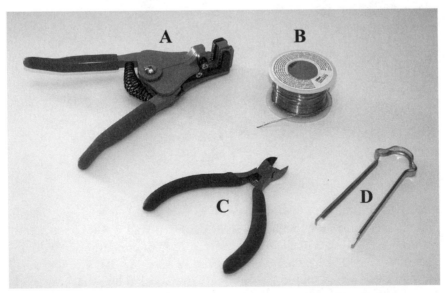

■ **FIGURE 1.9**   *Wire strippers, cutters, solder, and a chip-pulling device.*

thinnest solder that you can find. You will definitely need a chip-pulling tool (D) for removing the programmable integrated circuit (PIC) microcontroller chips from their sockets.

The PIC microcontrollers will be inserted and removed from the sockets on the controller boards many times, as the software is changed and the PIC is reprogrammed during experiments. An adjustable work stand, like the one shown in **Figure 1.10** (M), will be useful when soldering components to circuit boards, or holding wires when soldering them to header connectors. A utility knife (N) will also be helpful when cutting heat-shrink tubing or small parts.

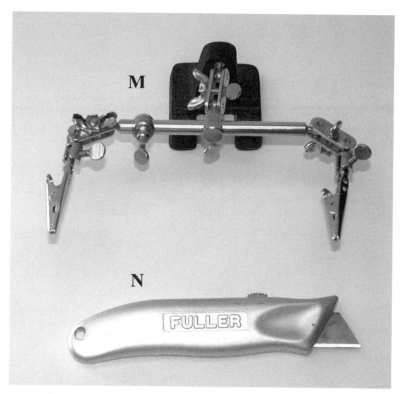

■ **FIGURE 1.10**   *Adjustable work stand and utility knife.*

A soldering iron, similar to the one shown in **Figure 1.11**, will be required when building the circuit boards for each project. An expensive soldering iron is not necessary, but the advantage to buying a good one is that the temperature can be set. A 15- to 25-watt pencil-style soldering iron will work and will help to protect delicate components from burning out.

An adjustable square (O) and a good ruler (P) will be required when measuring the cutting and drilling marks on the aluminum pieces that make up the parts for each

■ **FIGURE 1.11** *Soldering iron with adjustable temperature.*

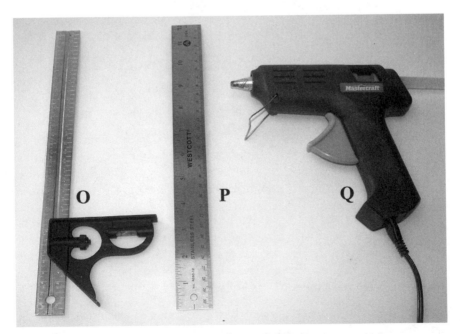

■ **FIGURE 1.12** *Adjustable square, ruler, and glue gun.*

project. You will need a hot glue gun (Q) and glue sticks at certain points in the construction. See **Figure 1.12**. A hammer (R), shown in **Figure 1.13**, will be needed for bending aluminum, along with a metal file (S) to smooth the edges of metal pieces after they have been cut or drilled. You may use a tube of quick-setting epoxy (T) to secure parts. Safety glasses (U) should be worn at all times when cutting and drilling metal or soldering.

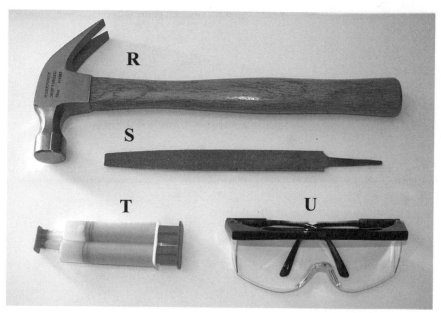

■ **FIGURE 1.13**   *Hammer, file, epoxy, and safety glasses.*

Test Equipment

To calibrate and troubleshoot the electronics, you will need a digital multimeter with frequency counting capabilities, similar to the Fluke 87 multimeter (**Figure 1.14**, left). When working with electronic circuits, a good multimeter is invaluable. The second multimeter in **Figure 1.14** (right) is manufactured by Circuit Test and measures capacitance, resistance, and inductance. It is nice to be able to measure the exact values of components when working on precise circuits, but not necessary in most cases. If you are winding your own transformers or chokes, the ability to measure inductance will be helpful. The specific use of the multimeter will be explained during the construction of the robot's electronics in later chapters.

If you are serious about electronics, an oscilloscope, like the one pictured in **Figure 1.15**, is a great investment. This is the Tektronix TDS 210 dual channel, digital real-time oscilloscope, with a 60-MHz bandwidth. The TDS 210 on my bench also has the RS-232, general purpose interface bus (GPIB), and Centronics port mod-

■ FIGURE 1.14   *Fluke and Circuit Test multimeters.*

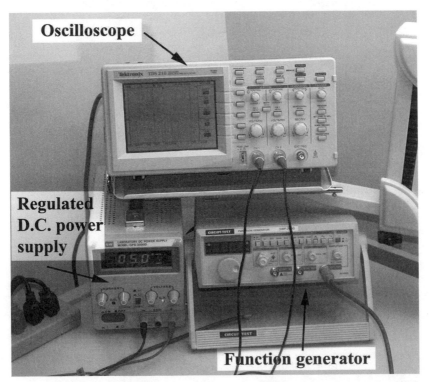

Oscilloscope

Regulated
D.C. power
supply

Function generator

■ FIGURE 1.15   *Oscilloscope, regulated DC power supply, and a function generator.*

ule added, so that a hard copy of waveforms can be printed, and it can communicate with a computer.

The great advantage to using an oscilloscope is the ability to visualize what is happening with a circuit. The new digital oscilloscopes also automatically calculate the frequency, period, mean, peak to peak, and true RMS of a waveform. You will probably need to use a regulated direct current (DC) power supply and a function generator quite often as well.

None of the equipment shown in **Figure 1.15** is required when building the projects in this book, but it will make your life as an electronics experimenter much easier. There is nothing more frustrating than finding out that a circuit you are working on is malfunctioning because of a dead battery or an oscillator calibrated to the wrong frequency. If you use a good power supply and oscilloscope when building and testing a circuit, the chance of these kinds of problems surfacing is much lower. I have found that if I am working late at night and start to encounter problems and make mistakes, it is best to shut my equipment down and get a good night's sleep. Sometimes the difference between frying an expensive chip or the circuit's working perfectly on the first try is just one misplaced component.

Construction Materials

Most of the projects in this book are constructed using aluminum and fasteners that are readily available at most hardware stores. Five sizes of aluminum will be used. The first stock measures 1/2-inch wide × 1/8-inch thick, and is usually bought in lengths of 4 feet or longer. Many of the parts for the projects are constructed from aluminum, with the dimensions as shown in **Figure 1.16**.

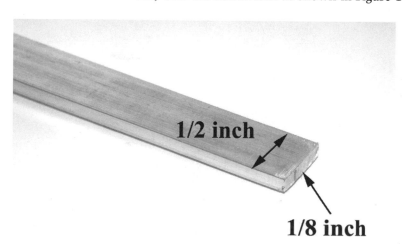

1/2 inch

1/8 inch

■ **FIGURE 1.16**   *1/2 × 1/8-inch aluminum stock.*

The second type of aluminum stock that will be used measures 1/4 inch × 1/4 inch, and is shown in **Figure 1.17**. It is usually bought in lengths of 4 feet or longer as well.

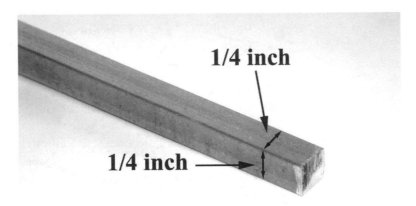

■ **FIGURE 1.17**   *Aluminum stock with 1/4 × 1/4-inch dimensions.*

The third kind of aluminum stock is 1/2-inch × 1/2-inch angle aluminum, and is 1/16-inch thick, as shown in **Figure 1.18**.

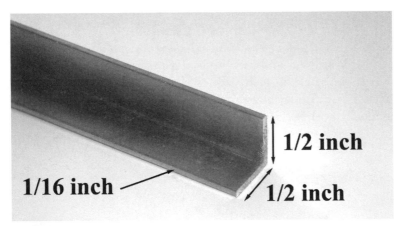

■ **FIGURE 1.18**   *1/2-inch angle aluminum.*

The fourth type is 1/16-inch thick flat aluminum, as shown in **Figure 1.19**, and it is usually bought in larger sheets. However, most metal suppliers will cut it down for you. This thickness of aluminum is great for cutting out custom parts, and it is easy to bend, making it ideal for the hobbyist experimenter. I buy all of my metal from The Metal Supermarket (www.metalsupermarkets.com) because its prices are much lower than buying metal at a hardware store. Its friendly staff is always helpful, and will cut the stock to any size. I usually ask the staff to cut the raw stock in half so that it will fit into the back seat of my car.

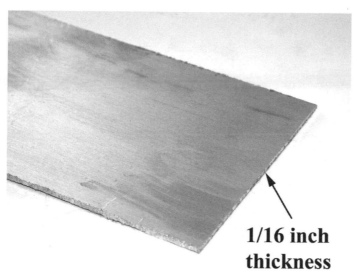

1/16 inch thickness

■ **FIGURE 1.19**   *1/16-inch thick flat aluminum.*

The fifth type of stock that will be needed is 3/4 × 3/4-inch angle aluminum. The fasteners that will be used are 6/32-inch diameter machine screws, nuts, lock washers, locking nuts, and nylon washers, as shown in **Figure 1.20**. Three different lengths of machine screws will be used: 1 inch, 3/4 inch, and 1/2 inch.

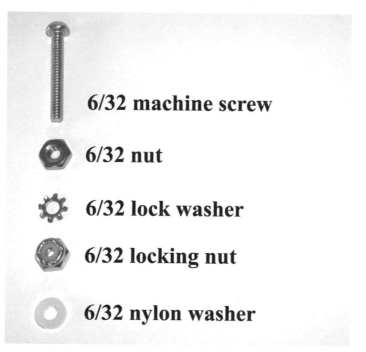

6/32 machine screw

6/32 nut

6/32 lock washer

6/32 locking nut

6/32 nylon washer

■ **FIGURE 1.20**   *6/32-inch diameter machine screw, lock washer, nuts, and nylon washer.*

Summary

Now that all the tools, test equipment, and materials have been covered, you should have a good idea about what will be necessary to build the mechanical parts for the projects in this book. In the next chapter, the fabrication of printed circuit boards will be described so that you can make your own professional-looking boards.

14

Printed Circuit Board Fabrication

Most of the robotics projects in this book will require some type of printed circuit boards (PCBs). The most efficient way of implementing the circuit designs in this book is to create PCBs. What's great about each project is that the finished PCB artwork is included, along with a parts placement diagram. All of the circuit boards and projects in this book have been built and tested to ensure that they function as described. If you decide not to fabricate PCBs, most of the circuits are simple enough to construct on standard perforated circuit board (holes spaced 0.1 inch on centers) using point-to-point wiring if you wish. I don't recommend this method because one misplaced or omitted wire can cause hours of frustration.

The easiest way to produce quality PCBs is by using the positive photo fabrication process. To fabricate the PCBs for each project, photocopy the PCB artwork onto a transparency. Make sure that the photocopy is the exact size of the original. For convenience, you can download the artwork files for each robot project from the Thinkbotics Web site, www.thinkbotics.com, and print the file onto a transparency using a laser or ink-jet printer with a minimum resolution of 600 dots per inch (dpi). **Figure 2.1** shows the artwork for a circuit board that has been printed onto transparency film using an ink-jet printer.

After successfully transferring the artwork to a transparency, use the following instructions to create a board. A 4×6-inch presensitized positive copper board is ideal for all of the projects presented in this book. When you place the transparency on the copper board, orient it exactly as shown in each chapter. Make any sensor boards that go with the particular project at the same time. M.G. Chemicals specializes in providing presensitized copper boards and all the chemistry needed to fabricate boards. Information on how to obtain all of the supplies can be found on its Web site: www.mgchemicals.com. **Figure 2.2** shows the developer, ferric

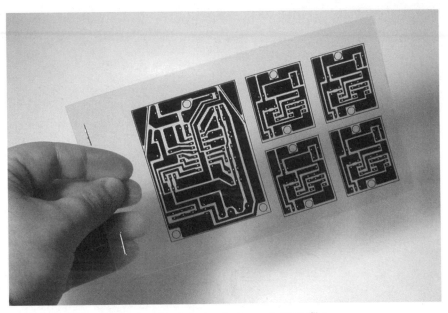

■ **FIGURE 2.1**   *PCB artwork printed onto transparency film.*

■ **FIGURE 2.2**   *Photo fabrication kit.*

chloride, and presensitized copper board that will be used for fabricating the circuit boards.

Follow the next six steps to make your own PCBs:

1. **Setting up**—Protect surrounding areas from developer and other splashes that may cause etching damage. Plastic is ideal for this. Work under safe light conditions. A 40-W incandescent bulb works well. Do not work under fluorescent light. Just prior to exposure, remove the white protective film from the presensitized board. Peel it back carefully.

2. **Exposing your board**—For best results, use the M.G. Chemicals cat. #416-X exposure kit. However, any inexpensive lamp fixture that will hold two or more 18-inch fluorescent tubes is suitable.

   Directions: Place the presensitized board, copper side toward the exposure source. Lay positive film artwork onto the presensitized copper side of the board and positioned as desired. Artwork should have been produced by a 600-dpi or better printer. If you don't have a 600-dpi or higher printer, then make two transparencies and lay them on top of each other. Make sure that the traces line up perfectly, then staple them together. A glass weight should then be used to cover the artwork, ensuring that no light will pass under the traces (approximately 3-mm glass thickness or greater works best). Use a 10-minute exposure time at a distance of 5 inches.

3. **Developing your board**—The development process removes any photoresist that was exposed through the film positive to ultraviolet light.

   Warning: The developer contains sodium hydroxide and is highly corrosive. Wear rubber gloves and eye protection while using it. Avoid contact with eyes and skin. Flush thoroughly with water for 15 minutes if it is splashed in eyes or on the skin.

   Directions: Using rubber gloves and eye protection, dilute one part M.G. cat. #418 developer with 10 parts tepid water (weaker is better than stronger). In a plastic tray, immerse the board, copper side up, into the developer. You will quickly see an image appear while you are lightly brushing the resist with a foam brush. This should be completed within one to two minutes. Immediately neutralize the development action by rinsing the board with water. The exposed resist must be removed from the board as soon as possible. When you are done with the developing stage, the only resist remaining will be covering what you want your circuit to be. The rest should be completely removed.

4. **Etching your board**—For best results, use the 416-E Professional Etching Process Kit or 416-ES Economy Etching Kit. The most popu-

lar etching matter is ferric chloride, M.G. cat. #415, an aqueous solution that dissolves most metals.

Warning: This solution is normally heated up during use, generating unpleasant and caustic vapors; adequate ventilation is very important. Use only glass or plastic containers. Keep out of reach of children. May cause burns or stain. Avoid contact with skin, eyes, or clothing. Store in a plastic container. Wear eye protection and rubber gloves.

If you use cold ferric chloride, it will take a long time to etch the board. To speed up the etching process, heat the solution. A simple way of doing this is to immerse the ferric chloride bottle or jug in hot water, adding or changing the water to keep it hot. Using thermostat-controlled crock pots or thermostatically controlled submersible heaters (glass enclosed, such as an aquarium heater) are effective ways to heat ferric chloride. An ideal etching temperature is 50°C (120°F). Be careful not to overheat the ferric chloride. The absolute maximum working temperature is about 57°C (135°F). The warmer your etch solution, the faster your boards will etch. Ferric chloride solution can be reused until it becomes saturated with copper. As the solution becomes more saturated, the etching time will increase. Agitation assists in removing unwanted copper faster. This can be accomplished by using air bubbles from two aquarium air wands with an aquarium air pump. Do not use an aquarium air stone. The etching process can be assisted by brushing the unwanted resist with a foam brush while the board is submerged in the ferric chloride. After the etching process is completed, wash the board thoroughly under running water. Do not remove the remaining resist protecting your circuit or image, as it protects the copper from oxidation. If you require it to be removed, use a solvent cleaner. **Figure 2.3** shows an etched board ready for drilling.

5. **Drilling and parts placement**—Use a 1/32-inch drill bit to drill all the component holes on the PCB. Drill the holes for larger components with a 3/64-inch bit where indicated. Drill any holes that will be used to mount the circuit board at this time. It is best to use a small drill press, like the one shown in **Figure 2.4**, rather than a hand drill, when working with circuit boards. This is to ensure that the holes are drilled straight and accurately.

6. **Soldering your board**—Removal of resist is not necessary when soldering components to your board. When you leave the resist on, your circuit is protected from oxidation. Tinplating your board is not necessary. In the soldering process, the heat disintegrates the resist underneath the solder, producing an excellent bond.

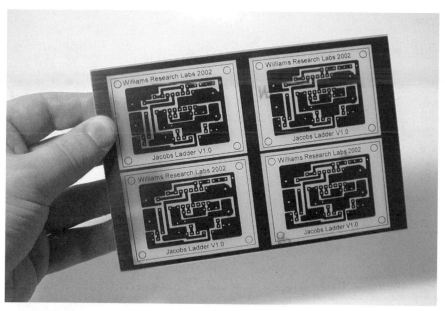

■ **FIGURE 2.3** *An etched board ready for drilling.*

■ **FIGURE 2.4** *A small drill press used to drill holes in a PCB.*

Summary

The ability to fabricate your own professional printed circuit boards using the photo fabrication process will make your projects much more reliable and efficient. The next chapter focuses on the PIC microcontroller and how it is programmed. Chapter 3 also covers the PicBasic Pro compiler, the EPIC hardware programmer, and the use of a development studio designed to speed up programming and debugging.

20

Microcontrollers and PIC Programming

Microcontrollers

The microcontroller is an entire computer on a single chip. The advantage of designing around a microcontroller is that a large amount of electronics needed for certain applications can be eliminated. This makes it the ideal device for use with mobile robots and other applications where computing power is needed. The microcontroller is popular because the chip can be reprogrammed easily to perform different functions and is very inexpensive. The microcontroller contains all the basic components that make up a computer. It contains a central processing unit (CPU), read-only memory, random-access memory (RAM), arithmetic logic unit, input and output lines, timers, serial and parallel ports, digital-to-analog converters, and analog-to-digital converters. The scope of this book is to describe the specifics of how the microcontroller can be used as the processors for the various humanoid robot projects that will be built.

PIC 16F84A MCU

Microchip technology has developed a line of reduced instruction set computer (RISC) microprocessors called the programmable interface controller (PIC). The PIC uses what is known as Harvard architecture. Harvard uses two memories and separate busses. The first memory is used to store the program, and the other is used to store data. The advantage of this design is that instructions can be fetched by the CPU at the same time that RAM is being accessed. This greatly speeds up execution time. The architecture commonly used for most computers today is known as Von Neumann architecture. This design uses the same memory for control and RAM storage, and slows down processing time.

For the first project in the book, we will use the PIC 16F84A, shown in **Figure 3.1**, as the processor for the robot arm controller circuit. This device can be reprogrammed because it uses flash read-only memory for program storage. This makes it ideal for experimenting because the chip does not need to be erased with an ultraviolet light source every time you tweak the code or try something new.

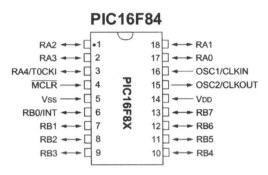

PIC16F84

■ **FIGURE 3.1** *Pinout of the PIC 16F84 microcontroller.*

The PIC 16F84A is an 18-pin device with an 8-bit data bus and registers. We will use a 4-MHz crystal for the clock speed. This is very fast for our application when you consider that it is running machine code at 4 million cycles per second. The PIC 16F84A is equipped with two input/output (I/O) ports, Ports A and B. Each port has two registers associated with it. The first register is the Tri State (TRIS) register. The value loaded into this register determines if the individual pins of the port are treated as inputs or outputs. The other register is the address of the port itself. Once the ports have been configured using the TRIS register, data can then be written or read to the port using the port register address.

Port B has eight I/O lines available and Port A has five I/O lines. For example, the first robot project in the book details the construction and programming of a robot arm. The robot arm's serial servo controller will use all eight I/O lines of Port B and all five lines of Port A, as shown in **Figure 3.2**.

Table 3.1 shows how the various pins of Ports A and B will be used as inputs and outputs to control the different functions of the robot arm. It is useful to have a list of the various I/Os connected to the ports when programming.

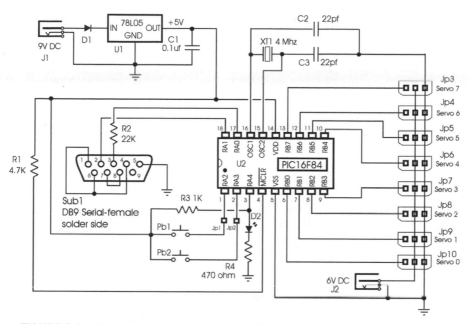

■ **FIGURE 3.2** *Robot arm serial servo controller board schematic.*

TABLE 3.1 *PIC 16F84A Port A and B Connection Table*

Port B	Configuration	Robot connection
RB0	Output	Base rotation servo
RB1	Output	Left shoulder servo
RB2	Output	Right shoulder servo
RB3	Output	Elbow servo
RB4	Output	Left wrist servo
RB5	Output	Right wrist servo
RB6	Output	Wrist rotation servo
RB7	Output	Gripper servo

Port A	Configuration	Robot connection
RA0	Input	Input from host computer
RA1	Output	Output to host computer
RA2	Input	Push button #1
RA3	Input	Push button #2
RA4	Output	Light-emitting diode

PicBasic Pro Compiler

MicroEngineering Labs develops the PicBasic Pro Compiler, shown in **Figure 3.3**. The programming language makes it quick and easy to program Microchip Technology's powerful PICmicro microcontrollers. It can be purchased from microEngineering Labs, whose Web site is www.melabs.com.

■ **FIGURE 3.3** *PicBasic Pro Compiler.*

The BASIC language is much easier to read and write than Microchip assembly language and will be used to program the robots in this book. The PicBasic Pro Compiler is "BASIC Stamp II like," and has most of the libraries and functions of both the BASIC Stamp I and II. Because it is a true compiler, programs execute much faster, and may be longer than their Stamp equivalents. One advantage of the PicBasic Pro Compiler is that it uses a real IF..THEN..ELSE..ENDIF, instead of the IF..THEN(GOTO) of the Stamps. These and other differences are spelled out in the PicBasic Pro manual. PicBasic Pro defaults to create files that run on a PIC 16F84A-04/P clocked at 4 MHz. Only a minimum of other parts are necessary: two 22pf capacitors for the 4-MHz crystal, a 4.7K pull-up resistor tied to the /MCLR pin, and a suitable 5-V power supply. Many PICmicros other than the 16F84, as well as oscillators of frequencies other than 4 MHz, may be used with the PicBasic Pro Compiler.

The PicBasic Pro Compiler produces code that may be programmed into a wide variety of PICmicro microcontrollers having from 8 to 84 pins and various on-chip features, including A/D converters, hardware timers, and serial ports. For general purpose PICmicro development using the PicBasic Pro Compiler, the PIC 16F84, 16F876, and 16F877 are the current PICmicros of choice. These microcontrollers use flash technology to allow rapid erasing and reprogramming to speed program debugging. With the click of the mouse in the programming software, the flash PICmicro can be instantly erased and then reprogrammed again and again. Other PICmicros in the 12C67x, 14C000, 16C55x, 16C6xx, 16C7xx, 16C9xx, 17Cxxx, and 18Cxxx series are either one-time programmable (OTP) or have a quartz window in the top (JW) to allow erasure by exposure to ultraviolet light for several minutes.

The PIC 16F84 and 16F87x devices also contain between 64 and 256 bytes of non-volatile data memory that can be used to store program data and other parameters, even when the power is turned off. This data area can be accessed simply by using the PicBasic Pro Compiler's READ and WRITE commands. Program code is always permanently stored in the PICmicro's code space, whether the power is on or off. By using a flash PICmicro for initial program testing, the debugging process may be sped along. Once the main routines of a program are operating satisfactorily, a PICmicro with more capabilities or expanded features of the compiler may be utilized.

Software Installation

The PicBasic Pro files are compressed into a self-extracting file on the diskette. They must be uncompressed to your hard drive before use. To uncompress the files, create a subdirectory on your hard drive called PBP or another name of your choosing by typing:

md PBP

at the DOS prompt. Change to the directory:

cd PBP

Assuming the distribution diskette is in drive a:, uncompress the files into the PBP subdirectory:

a:\pbpxxx -d

where xxx is the version number of the compiler on the disk. Don't forget the -d option on the end of the command. This ensures that the proper subdirectories within PBP are created.

Make sure that FILES and BUFFERS are set to at least 50 in your CONFIG.SYS file. Depending on how many FILES and BUFFERS are already in use by your system, allocating an even larger number may be necessary.

See the README.TXT file on the diskette for more information on uncompressing the files. Also, read the READ.ME file that is uncompressed to the PBP subdirectory on your hard drive for the latest PicBasic Pro Compiler information. **Table 3.2** lists the different PicBasic Pro Compiler statements that are available to the PICmicro software developer.

TABLE 3.2   *PicBasic Pro Statement Reference*

Statement	Description
@	Insert one line of assembly language code.
ADCIN	Read on-chip analog to digital converter.
ASM..ENDASM	Insert assembly language code section.
BRANCH	Computed GOTO (equiv. to ON..GOTO).
BRANCHL BRANCH	Out of page (long BRANCH).
BUTTON	Debounce and auto-repeat input on specified pin.
CALL	Call assembly language subroutine.
CLEAR	Zero all variables.
CLEARWDT	Clear (tickle) Watchdog Timer.
COUNT	Count number of pulses on a pin.
DATA	Define initial contents of on-chip EEPROM.
DEBUG	Asynchronous serial output to fixed pin and baud.
DEBUGIN	Asynchronous serial input from fixed pin and baud.
DISABLE	Disable ON DEBUG and ON INTERRUPT processing.
DISABLE DEBUG	Disable ON DEBUG processing.
DISABLE INTERRUPT	Disable ON INTERRUPT processing.
DTMFOUT	Produce touch-tones on a pin.
EEPROM	Define initial contents of on-chip EEPROM.
ENABLE	Enable ON DEBUG and ON INTERRUPT processing.
ENABLE DEBUG	Enable ON DEBUG processing.
ENABLE INTERRUPT	Enable ON INTERRUPT processing.
END FOR..NEXT	Stop execution and enter low power mode.
FOR..NEXT	Repeatedly execute statements.
FREQOUT	Produce up to 2 frequencies on a pin.
GOSUB	Call BASIC subroutine at specified label.
GOTO	Continue execution at specified label.
HIGH	Make pin output high.
HSERIN	Hardware asynchronous serial input.
HSEROUT	Hardware asynchronous serial output.
I2CREAD	Read bytes from I2C device.
I2CWRITE	Write bytes to I2C device.
IF..THEN..ELSE..ENDIF	Conditionally execute statements.

(continued on next page)

26

TABLE 3.2   *PicBasic Pro Statement Reference*

Statement	Description
INPUT	Make pin an input.
LCDIN	Read from LCD RAM.
LCDOUT	Display characters on LCD.
{LET}	Assign result of an expression to a variable.
LOOKDOWN	Search constant table for value.
LOOKDOWN2	Search constant/variable table for value.
LOOKUP	Fetch constant value from table.
LOOKUP2	Fetch constant/variable value from table.
LOW	Make pin output low.
NAP	Power down processor for short period of time.
ON DEBUG	Execute BASIC debug monitor.
ON INTERRUPT	Execute BASIC subroutine on an interrupt.
OUTPUT	Make pin an output.
PAUSE	Delay (1mSec resolution).
PAUSEUS	Delay (1uSec resolution).
PEEK	Read byte from register. (Do not use.)
POKE	Write byte to register. (Do not use.)
POT	Read potentiometer on specified pin.
PULSIN	Measure pulse width on a pin.
PULSOUT	Generate pulse to a pin.
PWM	Output pulse width modulated pulse train to pin.
RANDOM	Generate pseudo-random number.
RCTIME	Measure pulse width on a pin.
READ	Read byte from on-chip EEPROM.
READCODE	Read word from code memory.
RESUME	Continue execution after interrupt handling.
RETURN	Continue at statement following last GOSUB.
REVERSE	Make output pin an input or an input pin an output.
SERIN	Asynchronous serial input (BS1 style).
SERIN2	Asynchronous serial input (BS2 style).
SEROUT	Asynchronous serial output (BS1 style).
SEROUT2	Asynchronous serial output (BS2 style).
SHIFTIN	Synchronous serial input.
SHIFTOUT	Synchronous serial output.
SLEEP	Power down processor for a period of time.
SOUND	Generate tone or white-noise on specified pin.
SWAP	Exchange the values of two variables.
TOGGLE	Make pin output and toggle state.
WHILE..WEND	Execute statements while condition is true.
WRITE	Write byte to on-chip EEPROM.
WRITECODE	Write word to code memory.
XIN	X-10 input.
XOUT	X-10 output.

Compiling a Program

For operation of the PicBasic Pro Compiler, you will need a text editor or word processor for creation of your program source file, some sort of PICmicro programmer such as the EPIC Plus Pocket PICmicro Programmer, and the PicBasic Pro Compiler itself. Of course you also need a personal computer (PC) to run it.

Follow this sequence of events:

First, create the BASIC source file for the program, using your favorite text editor or word processor. If you don't have a favorite, DOS EDIT (included with MS-DOS) or Windows NOTEPAD (included with Windows and Windows 95/98) may be substituted. A great text editor called Ultraedit is available at: www.ultraedit.com. It is geared toward the software developer and does not add any undesirable formatting characters that will cause the compiler to error out. The source file name should (but is not required to) end with the extension .BAS. The text file that is created must be pure ASCII text. It must not contain any special codes that might be inserted by word processors for their own purposes. You are usually given the option of saving the file as pure DOS or ASCII text by most word processors.

Program 3.1 provides a good first test for programming a PIC and for testing the serial servo controller board when it is built in Chapter 5. You can type it in or download it from the author's Web site, www.thinkbotics.com, and follow the links for book software. The file is named ssc-test.bas and is listed in **Program 3.1**. The BASIC source file should be created in or moved to the same directory as the PBP.EXE file.

■ **PROGRAM 3.1**   *ssc-test.bas program listing.*

```
'_____
' Name     : ssc-test.bas
' Compiler : PicBasic Pro - microEngineering Labs
' Notes    : Program to test the serial servo controller board
'_____
TRISB = %00000000
TRISA = %00001101
DEFINE OSC 4
Button_1    VAR PORTA.2
Button_2    VAR PORTA.3
Led         VAR PORTA.4
I           VAR BYTE

LOW Led
Start:
   FOR I = 1 to 10
```

```
        HIGH Led
        PAUSE 200
        LOW Led
        PAUSE 200
    NEXT I
Loop:
    LOW Button_1
    LOW Button_2

    If Button_1 = 1 THEN HIGH Led

    IF Button_2 = 1 THEN LOW Led

GOTO Loop
END
```

Once you are satisfied that the program you have written will work flawlessly, you can execute the PicBasic Pro Compiler by entering PBP, followed by the name of your text file at a DOS prompt. For example, if the text file you created is named ssc-test.bas, at the DOS command prompt, enter:

PBP ssc-test.bas

The compiler will display an initialization (copyright) message and process your file. If it likes your file, it will create an assembler source code file (in this case, named ssc-test.asm) and automatically invoke its assembler to complete the task. If all goes well, the final PICmicro code file will be created (in this case, ssc-test.hex). If you have made the compiler unhappy, it will issue a string of errors that will need to be corrected in your BASIC source file before you try compilation again.

To help ensure that your original file is flawless, it is best to start by writing and testing a short piece of your program, rather than writing an entire 100,000-line monolith all at once and then trying to debug it from end to end.

If you don't tell it otherwise, the PicBasic Pro Compiler defaults to creating code for the PIC 16F84. To compile code for PICmicros other than the F84, simply use the -P command line option, described in the PicBasic manual, to specify a different target processor. For example, if you intend to run the above program, ssc-test.bas, on a PIC 16C74, compile it using the command:

PBP -p16c74 ssc-test.bas

An assembler source code file for ssc-test.bas is also generated. It is called ssc-test.asm. The assembler source code can be used as a guide if you want to explore assembly language programming because the listing shows the PicBasic Pro statement and the corresponding assembly code on the next line. The rest of the chapters describing software will not be addressing assembly code. We only need to be concerned with the PicBasic source code and the generated .HEX machine code, as listed in **Program 3.2**.

If you do not have the resources to buy the PicBasic Pro compiler, simply type the listings of the .HEX files into a text editor and save the file with the program name and .HEX extension. All program listings in the book can also be downloaded from www.thinkbotics.com. However, I recommend buying a copy of the compiler if you wish to experiment, change, or customize the programs. If you decide to continue with robotics and electronics, you will eventually need to buy a good compiler, such as PicBasic Pro, when working with microcontrollers.

■ **PROGRAM 3.2**   *ssc-test.hex program listing.*

```
:1000000028288F018E00FF308E07031C8F07031CEA
:10001000232803308D00DF300F2003288D01E83EB8
:100020008C008D09FC30031C18288C070318152838
:100030008C0764008D0F15280C181E288C1C222894
:10004000000022280800831303138312640008OOB1
:1000500083168601OD30850083120512831605126 2
:1000600083120130A40064000B30240203184628D8
:1000700005168316051 2C830831201200512831657
:10008000051 2C83083120120A40F332805118316EE
:1000900005118312851183168511 64008312051DD5
:1000A000552805168316051283126400851D5C28E9
:0E00B000051283160512831 2462863005D2890
:02400E00F53F7C
:00000001FF
```

Using the EPIC Programmer to Program the PIC

The two steps left are putting your compiled program into the PICmicro microcontroller and testing it. The PicBasic Pro Compiler generates standard 8-bit Merged Intel HEX (.HEX) files that may be used with any PICmicro Programmer, including the EPIC Plus Pocket PICmicro Programmer, shown in **Figure 3.4**.

PICmicros cannot be programmed with BASIC Stamp programming cables. An example of how a PICmicro is programmed using the EPIC Programmer with the DOS programming software follows. If Windows 95/98/NT is available, using the Windows version of EPIC Programmer software is recommended. Make sure that no PICmicros are installed in the EPIC Programmer programming socket or any

30

■ **FIGURE 3.4**   *EPIC Programmer by microEngineering Labs.*

attached adapters. Hook the EPIC Programmer to the PC parallel printer port using a DB25 male to DB25 female printer extension cable. Plug the alternating current (AC) adapter into the wall and then into the EPIC Programmer (or attach two fresh 9-V batteries to the programmer and connect the "Batt ON" jumper). The light-emitting diode (LED) on the EPIC Programmer may be on or off at this point. Do not insert a PICmicro into the programming socket when the LED is on, or before the programming software has been started.

Enter:

EPIC

at the DOS command prompt to start the programming software. The EPIC software should be run from a pure DOS session or from a full-screen DOS session under Windows or OS/2. (Running the DOS version under Windows is discouraged. Windows [all varieties] alters the system timing and plays with the port when you are not looking, which may cause programming errors.) The EPIC software will look around to find where the EPIC Programmer is attached and get it ready to program a PICmicro. If the EPIC Programmer is not found, check all the above connections and verify that no PICmicro or any adapter is connected to the programmer.

Typing:

EPIC /?

at the DOS command prompt will display a list of available options for the EPIC software.

Once the programming screen is displayed, use the mouse to click on Open file or press Alt-O on your keyboard. Use the mouse (or keyboard) to select ssc-test.hex or any other file you would like to program into the PICmicro from the dialog box. The file will load, and you should see a list of numbers in the window at the left. This is your program in PICmicro code. At the right of the screen is a display of the configuration information that will be programmed into the PICmicro. Verify that it is correct before proceeding. In general, the oscillator should be set to XT for a 4-MHz crystal, and the Watchdog Timer should be set to ON for PicBasic Pro programs. Most important, Code Protect must be OFF when programming any windowed (JW) PICmicro. You may not be able to erase a windowed PICmicro that has been code protected. **Figure 3.5** shows the EPIC MS-DOS interface. Insert a PIC 16F84A into the programming socket and click on Program or press Alt-P on the keyboard. The PICmicro will first be checked to make sure it is blank, and then your code will be programmed into it. If the PICmicro is not blank, and it is a flash device, you can simply choose to program over it without erasing first. Once the programming is complete and the LED is off, it is time to test your program.

■ **FIGURE 3.5** *EPIC graphic user interface.*

Testing the Serial Servo Controller Board

As an example of the steps that are followed after a PIC has been programmed, a description of the procedure from chapter 5 is outlined next. Later in Chapter 5, when the serial servo controller board is finished and the PIC 16F84A is programmed with the ssc-test.hex program, insert the PIC into the socket on the controller board. Place the PIC into the 18-pin IC socket, with the notch and pin 1 facing toward the two pushbutton switches, as shown in **Figure 3.6**. Power up the circuit by plugging the 9-V adapter into its plug. If all is well, the LED should flash on and off ten times. When the flashing is finished, use the left pushbutton to turn the LED on and the right button to turn the LED off. This ensures that the 16F84 was programmed and is functioning properly.

■ **FIGURE 3.6**   *PIC 16F84 inserted into IC socket on controller board.*

The rest of the serial servo controller board will be tested during Chapter 5. If nothing is happening when the power is switched on, try going through the process of programming the PIC again, and choose the verify option from the EPIC user interface. If the chip fails verification, check the RS-232 cable and power supply to the programmer. If that does not work, try using a different 16F84A chip. If there was no error when programming the PIC, insert it back into the controller board and make sure that pin 1 is facing toward the pushbuttons. Check to make sure that the 9-V adapter is plugged into the power outlet. Check the controller board for any missed components or cold solder connections.

MicroCode Studio Visual Integrated Development Environment

Mecanique's MicroCode Studio is a powerful, visual integrated development environment (IDE), with an in-circuit debugging (ICD) capability designed specifically for microEngineering Labs' PicBasic Pro Compiler. The MicroCode Studio user interface is shown in **Figure 3.7**. This studio makes programming PIC microcontrollers very easy with a one-button process of compiling, assembling, and programming. MicroCode Studio is free for noncommercial use and can be downloaded at www.mecanique.co.uk/code-studio/. It is not time-limited and does not have any nag screens. However, you can only use one ICD model with MicroCode Studio. MicroCode Studio is not copyright-free. If you wish to redistribute MicroCode Studio, or make it available on another server, you must contact Mecanique and obtain permission first.

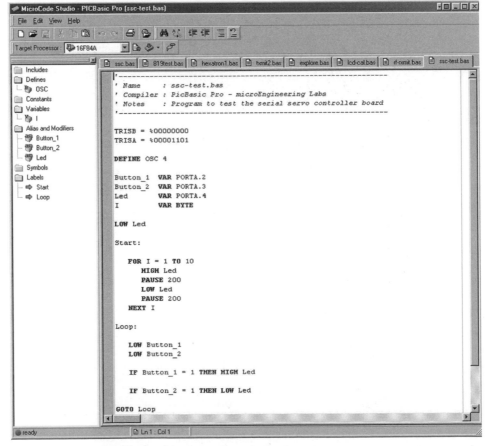

■ **FIGURE 3.7** *MicroCode Studio makes PIC programming easy.*

The main editor provides full syntax highlighting of your code, with context-sensitive keyword help and syntax hints. The code explorer allows you to automatically jump to include files, defines, constants, variables, aliases and modifiers, symbols, and labels that are contained within your source code. Full cut, copy, paste, and undo is provided, together with search and replace features. It also gives you the ability to identify and correct compilation and assembler errors. MicroCode Studio also lets you view serial output from your microcontroller. It includes keyword-based context-sensitive help, and it also supports MPASM and MPLAB.

It is easy to set up your compiler, assembler, and programmer options, or you can let MicroCode Studio do it for you with its built-in autosearch feature, as shown in **Figure 3.8**. MicroCode Studio has support for MPLAB-dependent programmers, such as PICStart Plus. Compilation and assembler errors can be easily identified and corrected using the error results window. Simply click on a compilation error and MicroCode Studio will automatically take you to the error line. MicroCode Studio even comes with a serial communications window, allowing you to debug and view serial output from your microcontroller.

■ **FIGURE 3.8** *Automatically setting up the compiler.*

With MicroCode Studio, you can start your preferred programming software from within the IDE. This enables you to compile and then program your microcontroller with just a few mouse clicks (or keyboard strokes, if you prefer). MicroCode Studio also supports MPLAB-dependent programmers.

Using a Programmer with MicroCode Studio

First, you need to tell MicroCode Studio which programmer you are using. Select VIEW...OPTIONS from the main menu bar, then select the PROGRAMMER tab, as shown in **Figure 3.9**.

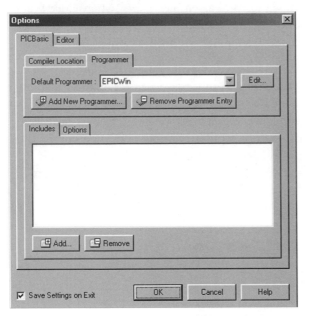

■ **FIGURE 3.9**   *Adding a new programmer.*

Next, select the Add New Programmer button. This will open the Add New Programmer wizard. Select the programmer you want MicroCode Studio to use, and then choose the Next button. MicroCode Studio will now search your computer until it locates the required executable. If your device uses MPLAB, you will be presented with two further screens, the select options and development mode screens. If your programmer is not in the list, you will need to create a custom programmer entry. Your programmer is now ready for use. When you press the Compile and Program button on the main toolbar, your PICBasic program is compiled and the programmer software is started. The hex filename and target device is automatically set in the programming software (if this feature is supported), ready for you to program the microcontroller, as shown in **Figure 3.10**.

MicroCode Studio In-circuit Debugger

The MicroCode Studio ICD enables you to execute a PICBasic Program on a host PIC microcontroller and view variable values, special function registers (SFRs), memory, and electronically erasable programmable memory (EEPROM) as the

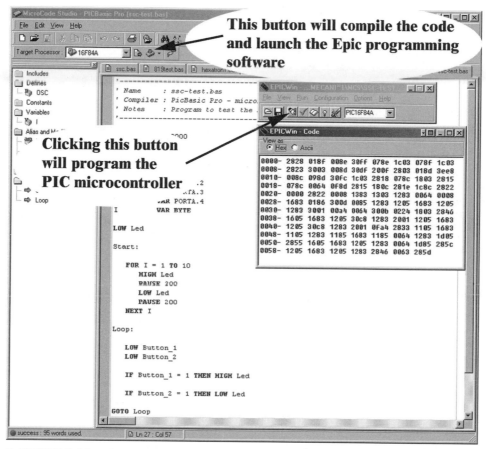

The following text is visible within the image:

This button will compile the code and launch the Epic programming software

Clicking this button will program the PIC microcontroller

```
' Name      : ssc-test.bas
' Compiler  : PicBasic Pro - micro
' Notes     : Program to test the
```

```
I          VAR PORTA.4
I          VAR BYTE

LOW Led

Start:

    FOR I = 1 TO 10
        HIGH Led
        PAUSE 200
        LOW Led
        PAUSE 200
    NEXT I

Loop:

    LOW Button_1
    LOW Button_2

    IF Button_1 = 1 THEN HIGH Led

    IF Button_2 = 1 THEN LOW Led

GOTO Loop
```

■ **FIGURE 3.10** *One button compile and programming using MicroCode Studio.*

program is running. Each line of source code is animated in the main editor window, showing you which program line is currently being executed by the host microcontroller. You can even toggle multiple breakpoints and step through your PICBasic code line by line. Using the MicroCode Studio ICD can really accelerate program development. It's also a lot of fun and a great tool for learning more about programming PIC microcontrollers.

Summary

Now that the concept of programming and compiling code for microcontrollers has been covered, it will be easy to program the microcontrollers for the projects in the following chapters. Using MicroCode Studio to create your source code, compiling the code, and programming PIC microcontrollers makes development much faster.

Humanoid Robotics Design Considerations

What makes us human? Is it possible to create an artificial human—an android? The human body, including the brain, is one of the most sophisticated biological machines in existence. The design of the human body is so complex that to build comparable machines, our technology will need to advance significantly. With our current understanding and technology, we can only attempt to mimic the form of the human body because it is impossible to recreate it at this time. It is no wonder that humans look to nature for insight and inspiration when designing machines. The idea of "reverse engineering" humans has fascinated humankind for a long time. The concepts of how the human body functions have proven to be extremely complex. For example, the human hand and wrist are very complicated devices for grasping and moving objects, but when implemented in machinery, they must be simplified to keep the mechanics and control systems within a reasonable level of complexity.

There are many reasons for wanting to create humanoid robots. Building robots that have a human form would allow those machines to take advantage of all the tools and equipment that have already been developed for humans. One main motivation for creating androids is the psychological aspect of human interaction with machines. We are much more comfortable communicating with machines that more closely resemble the human form, as opposed to machines that have an almost alien and sometimes frightening appearance. How often have you heard the phrase, "It's almost human!" when watching a robot do something interesting on a television show or at a science center? We humans quite often project our humanity onto machines and other life forms that resemble us. Another phrase heard quite often is, "It has a mind of it's own" when watching an automaton perform some entertaining task or a robot that senses and responds to its environment.

People would be much more comfortable interacting with machines that are designed to look like the human form. Now that automated banking machines have eliminated most of the human tellers, wouldn't it be nice to deal with the banking machines face-to-face and input your data without having to fumble with a card, cramped keyboard, and a small monitor? Imagine being able to walk up to a humanoid robot, have it access your banking information by facial recognition software, and then verify your identity with a retina scan. You would be able to talk to the machine in exactly the same way that you did with a live person.

In 1942, Isaac Asimov published his three laws of robotics in a short story called "Runaround," published by Street and Smith Publications. The three laws were stated as follows:

1. A robot may not injure a human being, or, through inaction, allow a human being to come to harm.
2. A robot must obey the orders given it by human beings except where such orders would conflict with the first Law.
3. A robot must protect its own existence as long as such protection does not conflict with the first or second law.

These laws could be incorporated into a set of rules defining robot morality, that is, if the robots being built are intended to respect human life. Most roboticists do not agree with Asimov's laws anymore. The first law disqualifies several important roles that humanoids are well suited to perform, such as soldier, police officer, and security guard. Much of the government funding for robotics, provided by the Defense Advanced Research Project Agency, is focused on military applications. The cruise missile is the perfect example of a fully autonomous robot that follows the first part of law 2 but, in doing so, breaks laws 1 and 3. The Predator robot, developed by the Central Intelligence Agency, is another example of a robot that kills with deadly precision by launching Hellfire missiles at its targets. It could be argued that these kinds of robots are now a necessity in the war against terrorists and rogue military nations that threaten national security.

To build an artificial person or humanoid, we must first consider what we are trying to construct. To answer that question, we need look no further than ourselves. The requirement specification would look somewhat like the following list.

1. The robot should have a more or less human form. It should have two legs, two arms, a torso, a human face, and a head. Its size should be roughly between 4 and 6 feet tall. The overall look of the robot should not stray considerably from what would be an acceptable human appearance. This is important because some people are afraid of robots that remind them of creatures like spiders, snakes, and lizards. I would consider one of the main motivations for creating humanoids

the psychological aspect of the acceptance of the machines by humans.

2. It should be able to communicate with humans in their native languages, without the use of an input device like a keyboard. The robot should at least respond to spoken commands, and it must be able to generate language by speech synthesis of some sort. It should also be able to convey simple emotions that correspond with the generated speech by using facial expressions.

3. It must be able to move from one location to another under its own free will or at the command of a human. While doing this, it must not harm any other objects or step on humans or pets.

4. It must be able to sense its environment and avoid obstacles and dangers that it might encounter along the way. A flight of stairs is not a problem for most healthy adults, but could be a catastrophic encounter for a humanoid robot or a frail person.

5. It must be able to pick up and carry objects in order to do some useful work such as vacuuming the carpet or cleaning the toilet. A robot arm and hand will be necessary to accomplish these tasks.

6. The humanoid should be able to learn from its own experience and then retain that information. It could then conceive its own strategies for dealing with those situations in the future.

7. The robot must possess some sort of intelligence and have the flexibility to accept training while adapting to the tasks that we wish it to perform. This would include solving simple problems that it might come across while carrying out the tasks.

8. The humanoid should understand and obey the basic principles of human social interaction. It should follow an acceptable code of behavior and possess a set of morals.

The preceding attributes are a very basic set of requirements. Many other components would need to be included to create a machine that would even come close to having the capabilities of a human. This book will tackle seven of the basic requirements that are listed and could be included in one machine. To give the specific details of building a humanoid from head to foot, a book of several thousand pages would be needed. We will start with building an anthropomorphic (human-like) robot arm and hand. Software to control the arm from a personal computer will be developed. The arm will have the ability to be remotely operated.

The next project will be the implementation of a speech recognition circuit that can be used to control the arm or other devices. The next project will be a robotic head and face that will generate speech and will express the emotional component of speech by facial expressions. The purpose of this project is to provide a humanoid robot with a means of communicating with humans in a more natural manner. The head will contain a speech synthesis circuit, a movable mouth, eyes, eyebrows, and facial muscles. When the head circuit receives a serial stream of text, it will convert this information to speech, along with mouth movement and the facial expression specified. The robot face will be capable of expressing emotions, such as happiness, sadness, anger, surprise, excitement, and others.

The final project will be the construction and programming of a bipedal humanoid walking robot. This project gives the reader the complete details needed to build and program a small-scale, fully autonomous walking humanoid robot, with the emphasis on bipedal walking.

Summary

There are many issues to consider when setting out to design and build a humanoid robot. Some of the considerations are more philosophical in nature than the mechanical, electronic, and programming achievements needed to create such machines. The projects presented in the next chapters will give you a good idea about what is involved in building some of the separate components that make up a humanoid robot.

42

5

Build Your Own Robot Arm, Gripper, and Serial Servo Controller

The first project in this book will detail the construction of the robot arm, wrist, and hand shown in **Figure 5.1**. This project also includes a controller board to interface the arm to a personal computer (PC) and to give the details needed to write the Visual Basic software used to control the arm. The arm controller board is designed to receive serial input commands so that it can be interfaced to any microcontroller or computer system. The controller board can also be used as a stand-alone device to control the arm from the onboard microcontroller.

A versatile arm is a component essential to a functional humanoid robot because it allows the robot to grasp and carry objects so that it can do useful work. In the field of robotics, there are five types of commonly used robot arm configurations, each named according to the combination of shoulder, elbow, and wrist joints. Each of the commonly used types employ simple revolute or prismatic joints. The five types are: the rectangular coordinate (this includes both floor and gantry mounts), spherical coordinate (polar), cylindrical coordinate, revolute coordinate, and the Self Compliant Automatic Robot Assembly (SCARA). Breaking from the regular categorization, two new classifications, serpentine and anthropomorphic, have arisen that utilize ball-and-socket or U-joints in their structures. The five principal types of robot arms are illustrated in **Figure 5.2**.

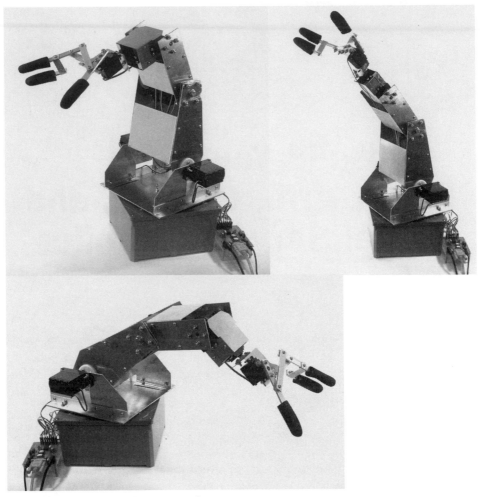

■ **FIGURE 5.1** *The robot arm project.*

Overview of the Robot Arm Project

The robot arm to be built in this chapter is the revolute type that closely resembles the human arm. The arm's rotating base is powered by a single large-scale servo that rotates the rest of the arm in a half-circle (180-degree) arc. Mounted to the base is an elevation joint, or shoulder, that can move the arm through 180 degrees, from horizontal to vertical on each side. The shoulder uses two large-scale servos, working together to provide the torque needed to lift the rest of the arm, as well as any object that it may be grasping. Attached to the shoulder piece is an elbow that can move through 180 degrees, also powered by a large-scale servo. The wrist is made up of three standard servos and can move through 180 degrees, from a

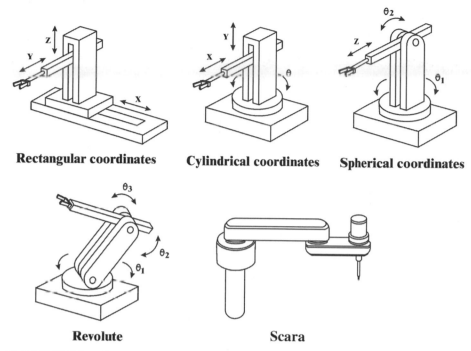

Rectangular coordinates   **Cylindrical coordinates**   **Spherical coordinates**

Revolute   **Scara**

■ **FIGURE 5.2**   *Five principal robot arm types.*

straight position to doubled back, as well as rotating the gripper clockwise and counterclockwise. Attached to the wrist is a three-fingered gripper that utilizes a unique design built around a single standard servo. The revolute geometry allows the robot arm to reach any point within a half-sphere, having the shape of an inverted bowl. The radius of the half-sphere should be the length of the arm when its shoulder, elbow, and wrist are straightened out.

The robot arm is controlled by a serial servo controller circuit board that will also be built in this chapter. The controller circuit board is based on Microchip's popular programmable integrated circuit (PIC) 16F84A flash programmable microcontroller, and it receives servo position commands from any device using a 2400-baud serial connection. This means that the arm can be used with any of the popular microcontroller systems available on the market or with a PC. The serial servo controller board will be connected to the serial port on a PC running the Microsoft Windows operating system. The robot arm control software that runs on the PC will be written in Visual Basic 6 and will be covered in the next chapter. The PIC 16F84A can also be programmed to run robot arm sequences independently. The controller will also be interfaced to a voice recognition circuit in Chapter 7.

Mechanical Construction

This project will begin with the mechanical construction and assembly of the robot arm. The parts needed to build the mechanical portion of the robot arm are listed in **Table 5.1**.

TABLE 5.1 *Parts List for the Robot Arm Mechanical Construction*

Parts	Quantity
6/32 × 1/2-inch machine screws	80
6/32 × 3/4-inch machine screws	8
6/32 × 1-inch machine screws	2
6/32 nylon washers	32
6/32 locking nuts	90
1/16-inch thick flat aluminum stock	6 foot × 7 foot sheet
1/2 × 1/8-inch aluminum stock	30 inches
3/4 × 3/4-inch angle aluminum	16 inches
1/2 × 1/2-inch angle aluminum	3 inches
1/4-inch diameter hollow aluminum tubing	2 inches
Plastic utility box: 8 inches wide × 8 inches in length × 4 inches deep—The Home Depot	1
Lazy Susan bearing platform: 6 inches × 6 inches	1
Cirrus 1/4-scale servo and hardware CS-600	4
Hitec standard servo and hardware HS-311	4
Servo connector extension wires: 24 inches in length	7
Bearing: 5/16-inch diameter	1
5/16 × 1 1/4-inch bolt	1
5/16 locking nut	1
Weather stripping foam	6 inches
Rubber glove	1

Fabrication of the Rotating Base and Shoulder

The rotating base, shoulder, and elbow of the robot arm are all powered by Cirrus CS-600 large-scale servos, like the one shown in **Figure 5.3**. The CS-600 servo provides 333 ounces/inches of torque, as compared to the 48.6 ounce/inch of torque delivered from a standard-sized servo. These servos are ideal for use in the arm joints that need to be able to lift a lot of weight. The 6 × 6-inch rotating "lazy Susan" bearing, shown in **Figure 5.4**, will be used to evenly distribute the weight of the robot arm when it is mounted to the base.

■ FIGURE 5.3 *Cirrus CS-600 large-scale servo.*

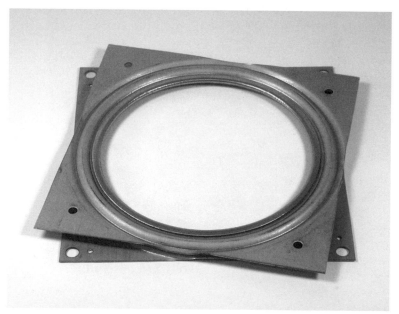

■ FIGURE 5.4 *Weight distribution bearing.*

Start by locating the plastic utility box. Drill a 1/4-inch hole in the bottom right-hand side of the box. The location of the hole, shown in **Figure 5.5**, does not need to be exact. Mark the center of the utility box lid with a pencil, and then mark an area 2 inches long × 1-1/4 inches wide on the lid, starting 1/2-inch above the center mark. Use a drill and a file to remove the plastic from this area until it is shaped like the hole shown in **Figure 5.6**. This area makes room to recess the top of the servo so that the base of the servo can be mounted flush to the lid. Position the raised area on the top of the servo into the recessed area so that the output shaft is at the center of the lid, and then mark the four mounting holes with a pencil. Drill the mounting holes in the lid with a 5/32-inch drill bit. Mount the servo from the bottom of the lid by inserting a 6/32 × 3/4-inch machine screw into each hole. Place three nylon washers between the lid and servo on each machine screw, and then secure in place with locking nuts, as shown in **Figure 5.7**.

Locate the rotating bearing and center it on top of the lid so that the space around it is even on all sides. Mark and drill the mounting holes on the outer corners of the bearing, and then secure in place with four 6/32 × 1/2-inch machine screw and locking nuts, as presented in **Figure 5.6**. This bearing will evenly distribute the weight of the arm across the base so that it can be easily turned by the base rotation servo.

Drill 1/4-inch hole

■ **FIGURE 5.5** *Access hole drilled in the side of the utility box.*

48

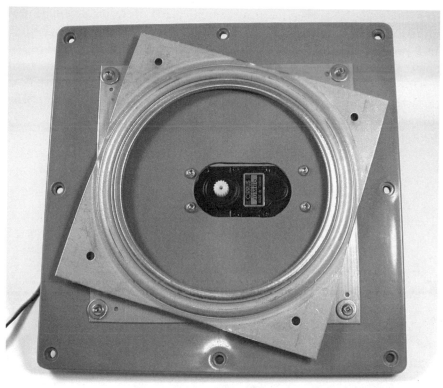

■ **FIGURE 5.6** *Servo and bearing assembly mounted to the utility box lid.*

■ **FIGURE 5.7** *Underside view of the base servo mounted to the lid.*

Build Your Own Robot Arm, Gripper, and Serial Servo Controller

To add stability to the base when the arm is mounted and operating, it will be necessary to add weight to the utility box before the lid is fastened in place. Fill the box with 8 pounds of weight, making sure to leave room for the servo when the lid is in place. I used scrap pieces of brass that were in one of my junk boxes. Feed the connector wire of the base servo through the 1/4-inch hole in the utility box, and then fasten the lid in place with the fastening screws that came with it. The base with the lid mounted in place is shown in **Figure 5.8**.

■ **FIGURE 5.8** *Completed servo, bearing, and lid attached to the utility box base.*

Cut a piece of 1/16-inch thick flat aluminum stock to a size of 8 inches × 8 inches. Drill the holes in the base piece (A), as marked in **Figure 5.9**. Cut two pieces (B and C) of 3/4-inch × 3/4-inch angle aluminum to a length of 8 inches each, and drill according to **Figure 5.10**. Locate each of the four 1 3/4-inch diameter servo horns that were supplied with the Cirrus jumbo servos. Drill four holes in each one at the locations shown in **Figure 5.11**, using a 5/32-inch bit.

Mount one of the jumbo servo horns (1 3/4-inch) to base piece A using four 6/32 × 1/2-inch machine screws and locking nuts, as exhibited in **Figure 5.12**. Flip base piece A over and attach mounting bracket pieces B and C to piece A using two 6/32 × 1/2-inch machine screws and locking nuts for each one. **Figure 5.13** shows the mounting brackets attached to the base piece.

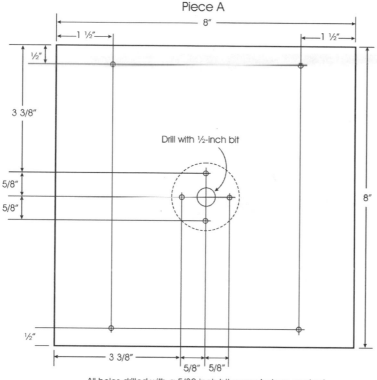

Piece A

Drill with ½-inch bit

All holes drilled with a 5/32-inch bit except where marked

■ **FIGURE 5.9** *Base cutting and drilling guide.*

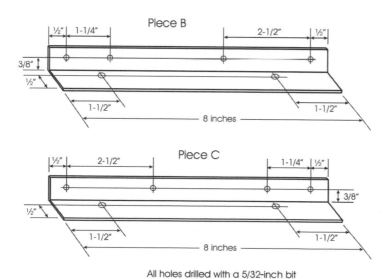

Piece B

Piece C

All holes drilled with a 5/32-inch bit

■ **FIGURE 5.10** *Cutting and drilling guide for mounting brackets.*

51

Build Your Own Robot Arm, Gripper, and Serial Servo Controller

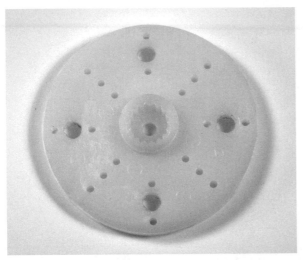

■ **FIGURE 5.11** *Large-scale servo horn (1 3/4-inch) drilling guide.*

■ **FIGURE 5.12** *Servo horn mounted to base piece A.*

Cut two pieces of 1/16-inch thick aluminum to a size of 3-5/8 inches × 4 inches, and then fashion two shoulder mounting brackets (D and E) according to the cutting, drilling, and bending dimensions shown in **Figure 5.14**. Note that for piece D, the aluminum is bent forward 90 degrees, and for piece E, it is bent backwards 90 degrees. Cut another two pieces of 1/16-inch thick aluminum to a size of 2-1/4 inches × 4 inches, and then fabricate two more shoulder mounting brackets (F and G) according to the cutting, drilling, and bending dimensions shown in **Figure 5.15**. Note that for piece F, the aluminum is bent forward 90 degrees, and for piece G, it is bent backwards 90 degrees. Be sure to file any rough edges.

■ **FIGURE 5.13**   Mounting brackets B and C attached to the base piece A.

Pieces D and E

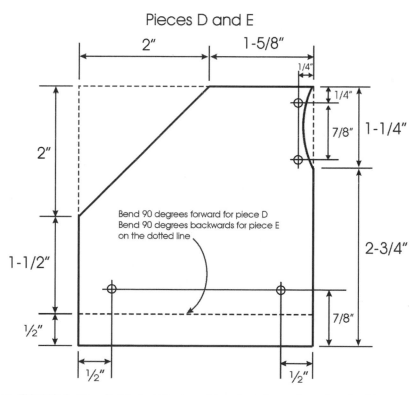

Bend 90 degrees forward for piece D
Bend 90 degrees backwards for piece E
on the dotted line

■ **FIGURE 5.14**   Cutting, drilling, and bending dimensions for shoulder mounting brackets D and E.

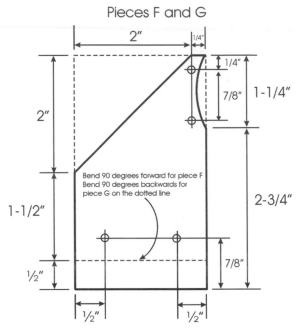

Pieces F and G

2"

1/4"

2"

1/4"

7/8" 1-1/4"

Bend 90 degrees forward for piece F
Bend 90 degrees backwards for
piece G on the dotted line

2-3/4"

1-1/2"

1/2"

7/8"

1/2"

1/2"

■ **FIGURE 5.15** *Cutting, drilling, and bending dimensions for shoulder mounting brackets F and G.*

Locate a Cirrus CS-600 servo and attach it to shoulder mounting brackets D and G, using four 6/32 × 1/2-inch machine screws and locking nuts, as shown in **Figure 5.16**. Locate another one of the jumbo servos and attach it to shoulder mounting brackets F and E, using four 6/32 × 1/2-inch machine screws and locking nuts, as shown in **Figure 5.17**.

■ **FIGURE 5.16** *Right shoulder elevation servo attached to mounting brackets D and G.*

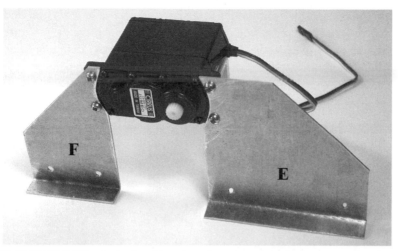

■ **FIGURE 5.17** *Left shoulder elevation servo attached to mounting brackets F and E.*

Attach the left shoulder assembly, made up of brackets E and F, to the base mounting bracket C using four 6/32 × 1/2-inch machine screws and locking nuts. Attach the right shoulder assembly, made up of brackets D and G, to the base mounting bracket B using four 6/32 × 1/2-inch machine screws and locking nuts. The entire shoulder drive assembly is shown in **Figure 5.18**.

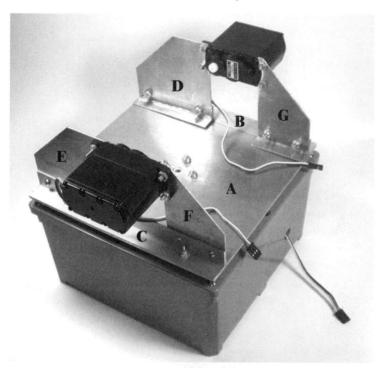

■ **FIGURE 5.18** *Shoulder elevation drive assembly attached to the rotating base.*

Cut two pieces of 1/16-inch thick aluminum to a size of 4 inches × 9 inches, and then cut and drill one of the pieces to the dimensions shown for piece H in **Figure 5.19**. Cut and drill the other piece to the dimension shown for piece I in **Figure 5.19**.

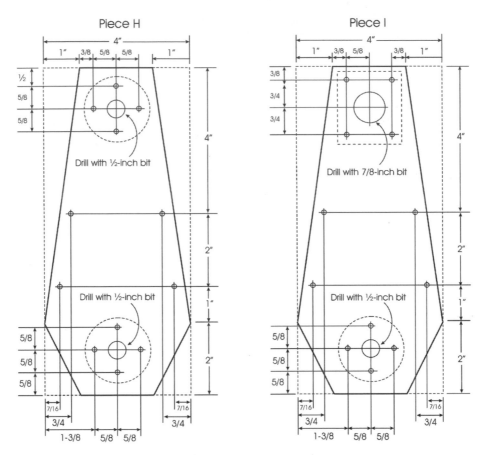

All holes drilled with a 5/32-inch bit except where marked

■ **FIGURE 5.19** *Cutting and drilling dimensions for shoulder pieces H and I.*

The next step is to fabricate a bracket to hold the elbow joint bearing in place. Cut a piece of 1/16-inch thick aluminum to a size of 1-3/4 inches × 2 inches, and then drill the holes in the positions shown in **Figure 5.20**. From a local hardware store, locate a bearing with a 5/16-inch inner diameter and dimensions of 1-1/4 inches wide × 3/4 inch in length, like the one shown in **Figure 5.21**. If you can't find this exact bearing, use any similar type with a 5/16-inch inner diameter. Mount the bearing to shoulder piece I by placing the wide end of the bearing flush with piece I, placing the bearing mount bracket over the bearing, and then securing it in place with four 6/32 × 3/4-inch machine screws and locking nuts. Fasten the locking nuts with enough pressure so that the mounting bracket piece J lies flat over the bearing and holds it securely in place. **Figure 5.22** shows the assembled elbow bearing and mount fastened to shoulder piece I.

Piece J

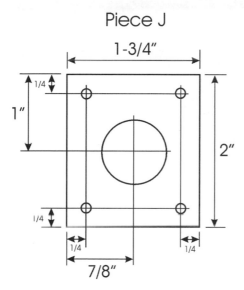

All holes drilled with a 5/32-inch bit except for the middle hole which uses a 7/8-inch bit

■ **FIGURE 5.20** *Elbow bearing mounting bracket.*

■ **FIGURE 5.21** *Elbow bearing.*

■ **FIGURE 5.22** *Assembled elbow bearing and mounting bracket attached to shoulder piece I.*

Attach a 1 3/4-inch diameter servo horn to shoulder piece I using four 6/32 × 1/2-inch machine screws and locking nuts on the opposite side of piece I that the bearing is mounted to, as shown in **Figure 5.23**. Attach another one of the 1 3/4-inch diameter servo horns to the top of shoulder piece H using four 6/32 × 1/2-inch machine screws and locking nuts. On the opposite side of piece H, to which the top servo horn was attached, mount the last 1 3/4-inch diameter servo horn to the bottom location. Use **Figure 5.23** as a guide.

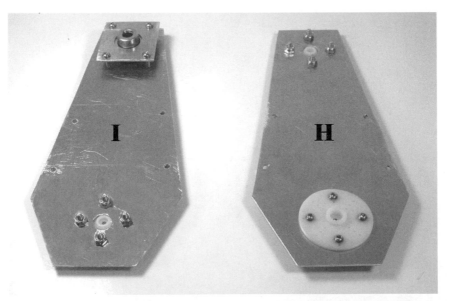

■ **FIGURE 5.23** *Servo horn mounting locations for shoulder pieces H and I.*

58

Fabricate two shoulder support pieces, K and L, using 1/16-inch thick aluminum cut to a size of 6-1/4 inches × 4 inches. Refer to **Figure 5.24** for the cutting, drilling, and bending dimensions needed to create the pieces. The two finished shoulder support pieces are shown in **Figure 5.25**. Attach the shoulder support pieces K and L to shoulder pieces H and I, in the positions shown in **Figure 5.26**, using eight 6/32 × 1/2-inch machine screws and locking nuts. The entire shoulder assembly is now complete and will be attached to the rotating base. Locate one of the extra servo horns that were supplied with the cirrus servos and attach it to one of the shoulder elevation servos. Turn the servo by hand all the way counterclockwise and then clockwise to determine its limits. Manually position the servo to its middle position. Perform the same procedure for the second shoulder servo. Mount the shoulder to the elevation servos so that the shoulder is in an upright vertical position and fasten it in place with the servo horn mounting screws. The completed shoulder and base assembly are shown in **Figure 5.27**.

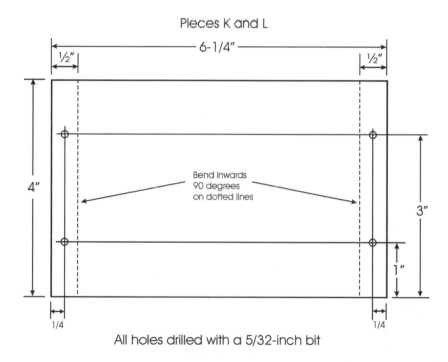

Pieces K and L

All holes drilled with a 5/32-inch bit

■ **FIGURE 5.24** *Cutting, drilling, and bending dimensions for shoulder support pieces K and L.*

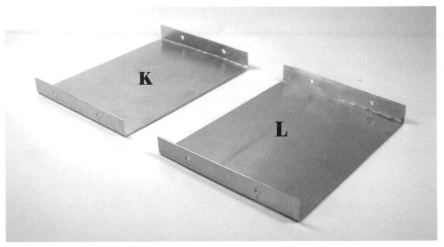

■ **FIGURE 5.25** *Completed shoulder support pieces K and L.*

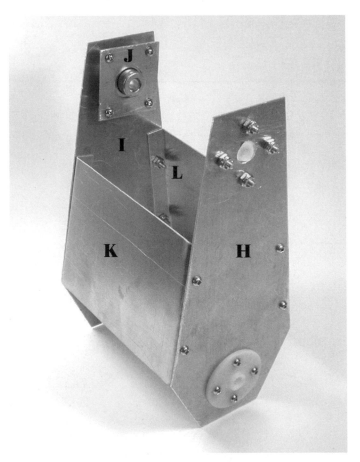

■ **FIGURE 5.26** *Completed shoulder assembly.*

■ **FIGURE 5.27**   *Shoulder assembly mounted to shoulder elevation servos.*

Constructing the Elbow and Wrist

Fabricate elbow pieces M and N using 1/16-inch thick aluminum. Refer to **Figure 5.28** for the cutting and drilling dimensions needed to create the pieces. Locate two of the cross-shaped servo horns that were supplied with the Hitec HS-311 servos and drill them with a 5/32-inch bit in the locations shown in **Figure 5.29**. Use **Figure 5.30** to position the servo horns on elbow pieces M and N, and then fasten each one in place with four 6/32 × 1/2-inch machine screws and locking nuts. Locate the last Cirrus CS-600 servo and fasten it to elbow piece N with the output shaft positioned closest to piece N, using two 6/32 × 1/2-inch machine screws and locking nuts, as shown in **Figure 5.30**. Locate an extra servo horn that was supplied with the Cirrus servo and attach it to the elbow servo. Turn the servo by hand all the way counterclockwise and then all the way clockwise to determine its limits. Manually position the servo to its middle position.

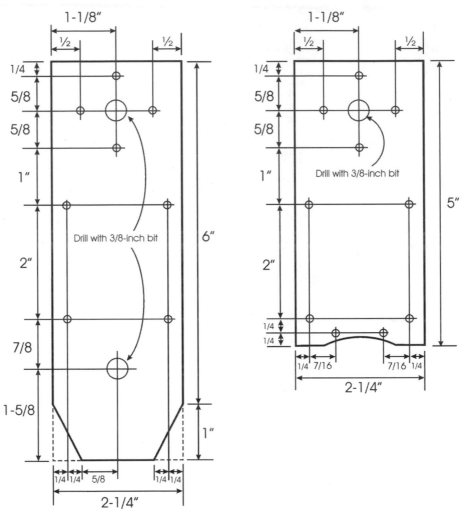

Piece M   Piece N

All holes drilled with a 5/32-inch bit except where marked

■ **FIGURE 5.28** *Cutting and drilling dimensions for elbow pieces M and N.*

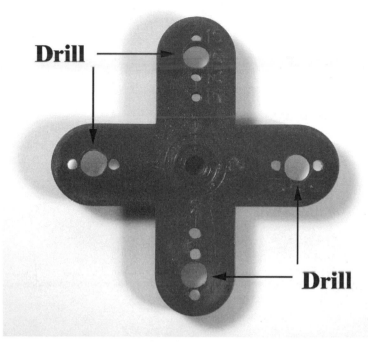

■ **FIGURE 5.29** *Servo horn drilling guide.*

■ **FIGURE 5.30** *Servo horns and CS-600 servo mounting guide.*

Fabricate two elbow support pieces, O and P, using 1/16-inch thick aluminum cut to a size of 4-7/8 inches × 3 inches. Refer to **Figure 5.31** for the cutting, drilling, and bending dimensions needed to create the pieces. The two finished shoulder support pieces are shown in **Figure 5.32**. Attach the elbow support pieces O and P to elbow pieces M and N, in the positions shown in **Figure 5.33**, using eight 6/32 × 1/2-inch machine screws and locking nuts.

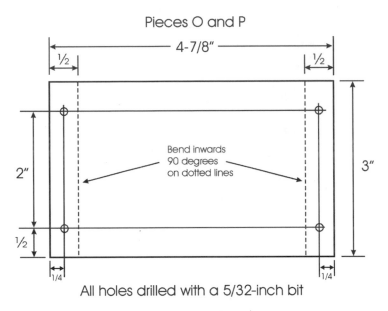

Pieces O and P

Bend inwards
90 degrees
on dotted lines

All holes drilled with a 5/32-inch bit

■ **FIGURE 5.31** *Cutting, drilling, and bending dimensions for elbow support pieces O and P.*

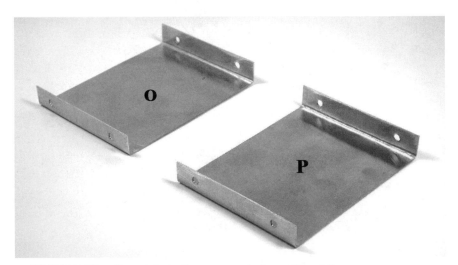

■ **FIGURE 5.32** *Completed elbow support pieces O and P.*

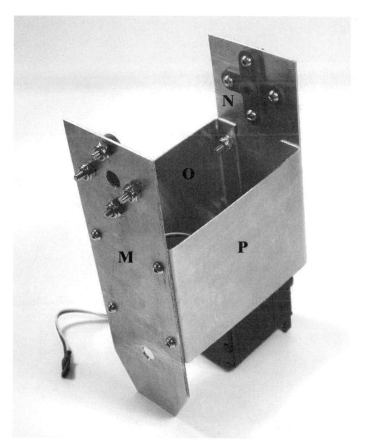

■ **FIGURE 5.33** *Completed elbow assembly.*

The next part of the arm to be built is the wrist. The wrist is made up of three standard Hitec servos and will allow the gripper to move horizontally and to rotate 180 degrees. Cut two pieces of 1/16-inch thick aluminum to a size of 1-3/4 inches × 1-7/8 inches. Drill and bend the two pieces, Q and R, according to the dimensions shown in **Figure 5.34**. Use **Figure 5.35** as a guide to attach each of the three Hitec servos to pieces Q and R, using eight 6/32 × 1/2-inch machine screws and locking nuts. The next two pieces of aluminum to be constructed are the top wrist covers. The wrist covers are fabricated from 1/16-inch thick aluminum cut to a size of 3-1/2 inches × 3 inches. Use **Figure 5.36** to cut, drill, and bend the top wrist cover pieces S and T. **Figure 5.37** shows finished wrist pieces S and T. To complete the wrist, attach pieces S and T to pieces Q and R with four 6/32 × 1/2-inch machine screws and locking nuts, as pictured in **Figure 5.38**.

Pieces Q and R

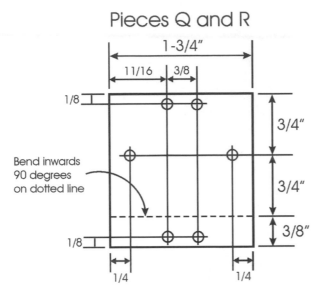

1-3/4"

11/16 3/8

1/8

3/4"

Bend inwards
90 degrees
on dotted line

3/4"

3/8"

1/8

1/4   1/4

All holes drilled with a 5/32-inch bit

■ **FIGURE 5.34** *Cutting, drilling, and bending dimensions for wrist pieces Q and R.*

■ **FIGURE 5.35** *Servos attached to pieces Q and R.*

Pieces S and T

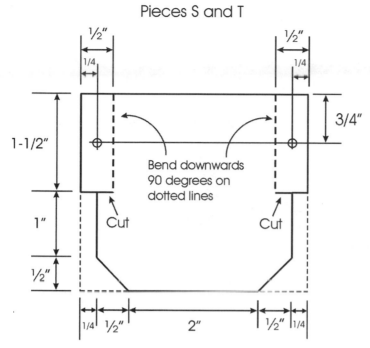

1/2"
1/4

1/2"
1/4

3/4"

1-1/2"

Bend downwards
90 degrees on
dotted lines

Cut   Cut

1"

1/2"

1/4 1/2"   2"   1/2" 1/4

■ FIGURE 5.36 *Cutting, drilling, and bending dimensions for wrist cover pieces S and T.*

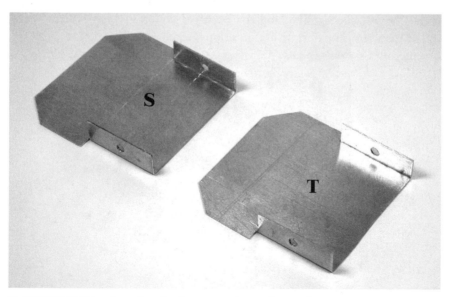

S

T

■ FIGURE 5.37 *Completed wrist cover pieces S and T.*

■ **FIGURE 5.38** *Completed wrist assembly.*

Locate an extra servo horn supplied with the Hitec CS-600 servo and attach it to one of the wrist servos. Turn the servo by hand all the way counterclockwise and then all the way clockwise to determine its limits. Manually position the servo to its middle position. Perform the same procedure for the other two servos. Mount the wrist on the elbow piece by attaching the wrist servo output shafts to the servo horns so that the wrist is in a horizontal position. Fasten the wrist in place with the servo horn mounting screws. **Figure 5.39** shows the wrist fastened to the elbow.

■ **FIGURE 5.39** *Wrist attached to the elbow section.*

The next step is to secure the elbow and wrist assembly to the shoulder section. With the elbow piece positioned vertically to the shoulder piece, attach the servo output shaft of the elbow to the servo horn on the end of the shoulder section. Secure the servo in place with the servo horn mounting screw. Push the 5/16 × 1 1/4-inch bolt through the hole in the bearing on the shoulder section and through the hole in the elbow piece. Secure the elbow piece tightly in place to the bearing with a 5/16 locking nut, as illustrated in **Figure 5.40**.

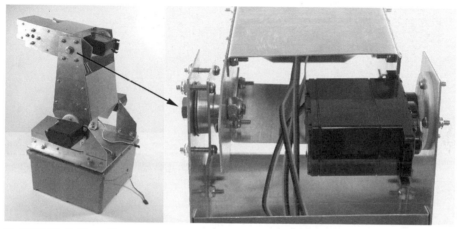

■ **FIGURE 5.40** *Elbow assembly attached to the shoulder section.*

Constructing the Robotic Gripper

Now that the mechanical construction of the arm is complete, a gripper will be built to give the arm the capability to manipulate objects. The gripper that will be built uses a single Hitec HS-311 standard servo. This design drives one of the fingers directly from the servo and the other two fingers through a direct mechanical linkage. The gripper is capable of a 10-inch grasp and is shown in **Figure 5.41**.

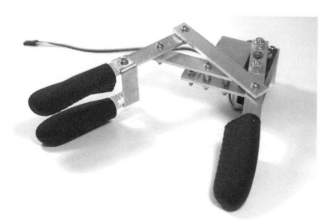

■ **FIGURE 5.41** *Robotic gripper.*

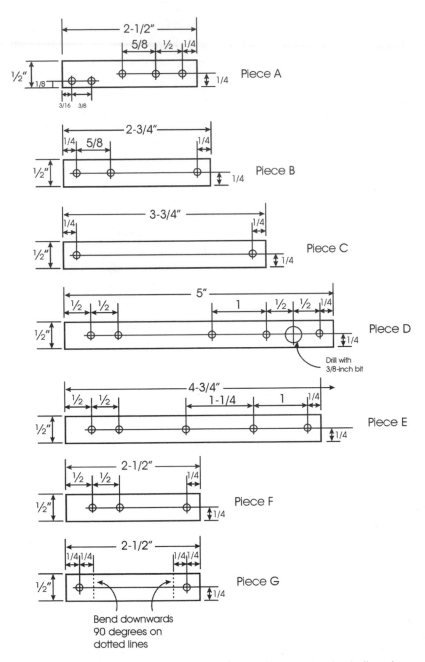

All holes drilled with a 5/32-inch bit except where marked otherwise

■ **FIGURE 5.42** *Cutting, drilling, and bending dimensions for gripper parts.*

Start by cutting and drilling the seven aluminum pieces, A, B, C, D, E, F, and G, according to the dimensions shown in **Figure 5.42**. Use standard 1/2 × 1/8-inch aluminum stock for all of the pieces except piece G, which is constructed using 1/16-inch flat stock so that the part can be easily bent. Next, cut two standoff pieces (H and I) from 1/4-inch diameter aluminum tubing and three finger pieces (J, K, and L) from 1/2-inch angle aluminum, as detailed in **Figure 5.43**. Fabricate the gripper base piece from 1/16-inch thick flat aluminum from a piece 2-1/2 inches × 4 inches in size. Cut, drill, and bend the piece according to the mechanical drawing in **Figure 5.44**.

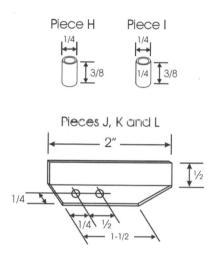

All holes drilled with a 5/32-inch bit

■ **FIGURE 5.43**   *Cutting and drilling guide for standoff and finger pieces.*

Now that all of the parts for the gripper have been fabricated, they will be assembled. Start by locating a Hitec HS-311 servo and attaching it to piece A, with two 6/32 × 1/2-inch machine screws and locking nuts, in the position shown in **Figure 5.45**. Attach the gripper base to the servo and piece A with three 6/32 × 1/2-inch machine screws and locking nuts, as shown in **Figure 5.45**. Use **Figure 5.46** to mount piece B to piece A through the two standoff pieces H and I using two 6/32 × 1-inch machine screws and locking nuts.

Using **Figure 5.47** as a guide, cut four pieces off of the servo horn using a hack saw or a pair of side cutters (as shown in **Figure 5.49**), and then file the edges smooth. The two pieces to be cut off are marked with the numbers 2 and 4 in the plastic, and are slightly smaller in width than the other two. Drill two holes, as indicated in **Figure 5.48**, using a 5/32-inch drill bit. The holes are drilled through the existing smaller holes; the third hole from the end of the horn on each side is marked as 13 in the plastic. When finished, the cut and drilled servo mount should look like the one shown in **Figure 5.50**.

Piece M

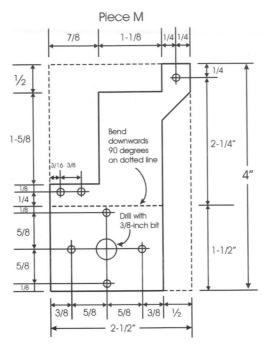

All holes drilled with a 5/32-inch bit

■ FIGURE 5.44 *Cutting, drilling, and bending dimensions for the gripper base.*

■ FIGURE 5.45 *Gripper base assembly diagram.*

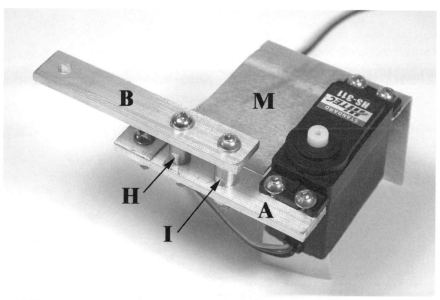

■ **FIGURE 5.46**   *Gripper assembly diagram.*

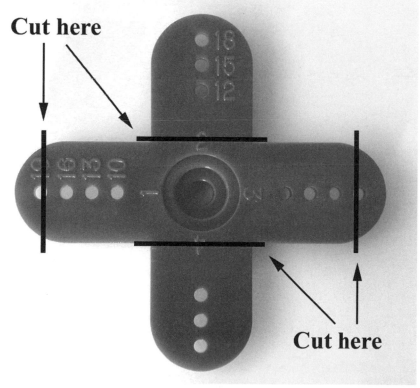

Cut here

Cut here

■ **FIGURE 5.47**   *Servo horn modification cutting guide.*

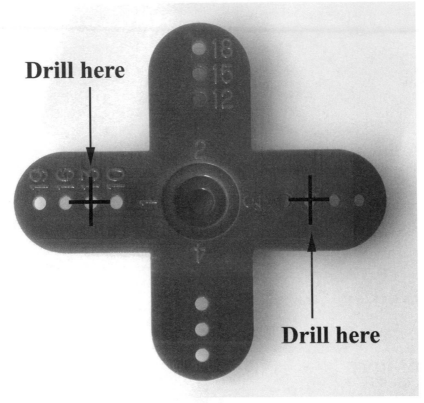

Drill here

Drill here

■ **FIGURE 5.48** *Servo horn modification drilling guide.*

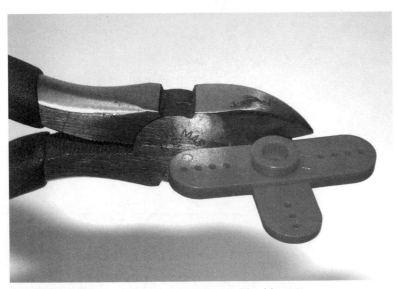

■ **FIGURE 5.49** *Cutting the servo horn with side cutters.*

■ **FIGURE 5.50** *Completed servo mount.*

Cut three pieces of weather stripping foam to a length of 2 inches each. Peel the coating off of the side with the sticky tape, and then stick each one to the square side of pieces J, K, and L. Piece J is shown in **Figure 5.51** with the weather stripping foam added. This will add a pliable cushion to the fingers when the gripper is picking up objects. When the foam has been added to the finger pieces, mount piece J to piece D using two 6/32 × 1/2-inch machine screws and locking nuts, as shown in **Figure 5.51**. Mount the modified servo horn to piece D using two 6/32 × 1/2-inch machine screws and locking nuts, as oriented in **Figure 5.51**. Next, join mechanical linkage piece C to piece D by adding a 6/32-inch nylon washer between the two pieces and then fastening it with a 6/32 × 1/2-inch machine screw and locking nut. Tighten the nut with enough force to allow the pieces to move freely

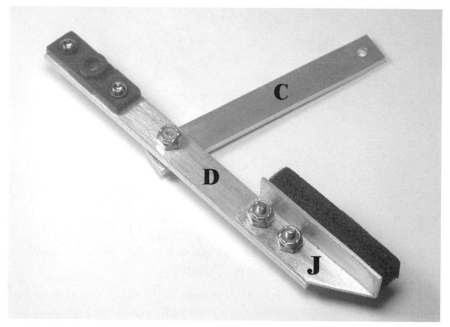

■ **FIGURE 5.51** *Servo mount finger assembly.*

without any friction, but enough to hold them in place. Use **Figure 5.51** as a guide when attaching the pieces. When this part is complete, attach the servo horn to the servo output shaft after the shaft has been manually rotated to its center position. Secure in place with the servo horn mounting screw. Assemble the rest of the gripper finger parts by following the parts placement diagram in **Figure 5.52**. You will need six 6/32 × 1/2-inch machine screws and locking nuts. Using **Figure 5.53** as a guide, attach piece E to piece B, separating the pieces with two 6/32-inch nylon washers. Secure in place with a 6/32 × 1/2-inch machine screw and locking nut. Tighten the nut with enough force to allow piece E to move freely without any friction, but enough force to hold it in place. Attach piece C to piece E, separating the pieces with a 6/32-inch nylon washer. Secure in place with a 6/32 × 1/2-inch machine screw and locking nut. Tighten the nut with enough force to allow the pieces to move freely without any friction, but enough force to hold them in place. The next step is to add the wrist servo horn to the gripper base. **Figure 5.54** shows the position where the wrist servo horn is mounted to the base using four 6/32 × 1/2-inch machine screws and locking nuts.

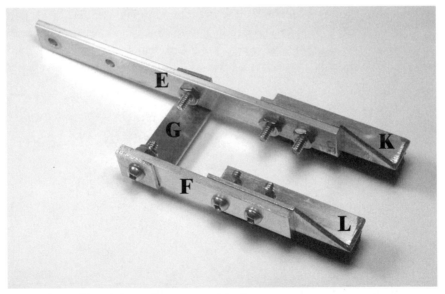

■ **FIGURE 5.52**   *Gripper finger assembly diagram.*

Cut three fingers off of a rubber glove to a length of 3 inches. Slip a rubber glove finger end onto the thumb and two fingers of the gripper.

The final step to completing the mechanical portion of the robot arm is to mount the gripper to the wrist. Position the gripper on the wrist's rotational servo, as pictured in **Figure 5.55**, and secure it in place with the servo horn mounting screw.

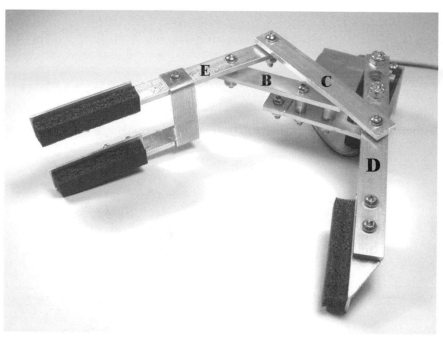

■ **FIGURE 5.53** *Gripper mechanical assembly diagram.*

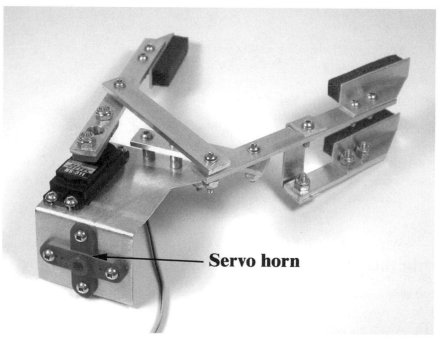

■ **FIGURE 5.54** *Wrist servo horn mount.*

■ **FIGURE 5.55** *Completed mechanical portion of the robot arm.*

Adding Servo Extension Wires

Extension connector wires will need to be added to each of the servos in preparation for connecting each one to the robot arm serial servo controller circuit board. Plug an extension wire into each of the servos and feed each wire down through the inside structure of the robot arm to the back of the base, where the base rotation servo wire comes out through the hole in the back.

Constructing the Serial Servo Controller Circuit Board

The serial servo controller board will be used to control the robot arm servos via a serial connection to a PC or directly from the onboard PIC 16F84 microcontroller. The schematic for the serial servo controller board is shown in **Figure 5.56**. **Table 5.2** lists all of the parts necessary to create the serial servo controller board.

Table 5.2 *Parts List for the Serial Servo Controller Board.*

Part	Quantity	Description
Semiconductors		
U1	1	78L05 5V regulator
U2	1	PIC 16F84 flash microcontroller mounted in socket
D1	1	1N4004 diode
D2	1	Red light-emitting diode
Resistors		
R1	1	4.7 KΩ 1/4-W resistor
R2	1	22 KΩ 1/4-W resistor
R3	1	1 KΩ 1/4-W resistor
R4	1	470Ω 1/4-W resistor
Capacitors		
C1	1	0.1 µf capacitor
C2, C3	2	22 pf
Miscellaneous		
JP1–JP2	2	1 post header connector—2.5 mm spacing
JP3–JP10	8	3 post header connector—2.5 mm spacing
PB1, PB2	2	Momentary contact push buttons
SUB1	1	DB9—female 9 pin serial connector
XT1	1	4 MHz crystal
J1, J2	2	Power jacks
Standoffs with mounting screws	4	1-1/4 inch in length
9-V DC adapter	1	250 ma
6-V DC adapter	1	2.5 amps
Serial cable	1	4 feet in length
Printed circuit board	1	See details in chapter

The serial servo controller circuit is designed around the Microchip PIC 16F84 microcontroller, clocked at 4 MHz. The circuit accepts information from a serial connection via the DB9 serial connector. We will be connecting the serial servo controller board to the serial port of a PC running serial servo control software that will be developed during the next chapter. The serial port of a PC uses the RS-232 communications protocol. While single-chip RS-232–level converters are common and inexpensive, the excellent input/output (I/O) specifications of the PICmicro microcontroller unit allows our application to run without level converters. Rather, inverted input (N300..N2400) can be used in conjunction with a current-limiting resistor. Eight of the PIC 16F84 I/O pins are connected to the robot arm servos.

The PIC is programmed to listen for any incoming serial communication from the host computer, and then set the servos to the positions received. The PIC is programmed to constantly update the servos with their positional information so that

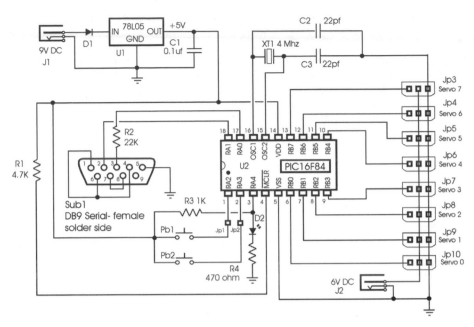

■ FIGURE 5.56   *Schematic for the serial servo controller circuit board.*

the servos will hold their positions. An explanation of how remote control (R/C) servos work and how they are controlled will make it clear why this is necessary.

The R/C servo is a geared, direct current (DC) motor with a built-in positional feedback control circuit. This makes it ideal for use with small robots because the experimenter does not have to worry about motor control electronics.

A potentiometer is attached to the shaft of the motor and rotates along with it. For each position of the motor shaft and potentiometer, a unique voltage is produced. The input control signal is a variable-width pulse between 1 and 2 milliseconds (ms), delivered at a frequency between 50 and 60 Hz, which the servo internally converts to a corresponding voltage. The servo feedback circuit constantly compares the potentiometer signal to the input control signal provided by the microcontroller. The internal comparator moves the motor shaft and potentiometer either forward or in reverse, until the two signals are the same. Because of the feedback control circuit, the rotor can be accurately positioned and will maintain the position as long as the input control signal is applied. The shaft of the motor can be positioned through 180 degrees of rotation, depending on the width of the input signal.

The PicBasic Pro language makes servo control with a PIC microcontroller easy, using a command called Pulsout. The syntax is Pulsout Pin, Period. A pulse is generated on Pin of specified Period. Toggling the pin twice generates the pulse; thus, the initial state of the pin determines the polarity of the pulse. Pin is automatically made an output. Pin may be a constant, 0–15, or a variable that contains a number between 0 and 15 (e.g., B0) or a pin name (e.g., PORTA.0).

The resolution of Pulsout is dependent on the oscillator frequency. Because we are using a 4-MHz oscillator, the Period of the generated pulse will be in 10 μs increments. To send a pulse to port B on pin 7 that is 1.4 ms long (at 4 MHz, 10 μs × 140 = 1400 μs or 1.4 ms), the command would be: Pulsout PortB.7,140. To illustrate the kind of signal being produced by the microcontroller, see **Figure 5.57**. The oscilloscope trace for channel 1 was generated with the Pulsout command configured to produce a 1.4-ms pulse at 55.68 Hz, and the trace for channel 2 was configured for a 6-ms pulse, also at 55.68 Hz.

The serial servo controller circuit board also includes a 5-V DC regulator to provide power to the PIC 16F84 microcontroller. Two push-button switches on the board can be programmed for mode selection, to run a sequence or any other function that the user desires. A light-emitting diode can also be user programmed. The servos have their own 6-V power supply so that any noise generated by the motors does not interfere with the microcontroller.

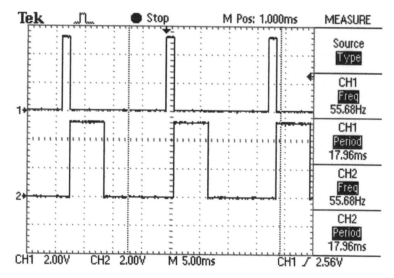

■ **FIGURE 5.57**   *Oscilloscope display of a 1.4-ms and 6.0-ms pulse train.*

Creating the Serial Servo Controller Printed Circuit Board

To fabricate the serial servo controller printed circuit board (PCB), photocopy the artwork in **Figure 5.58** onto a transparency. Make sure that the photocopy is the exact size of the original. For convenience, you can download the file from the author's Web site, www.thinkbotics.com, and simply print the file onto a transparency using a laser or ink-jet printer with a minimum resolution of 600 dpi. After the artwork has been successfully transferred to a transparency, use the techniques outlined in Chapter 2 to create a board. A 4 × 6-inch presensitized positive copper board is ideal. When you place the transparency on the copper board, it should be oriented exactly the same as in **Figure 5.58**.

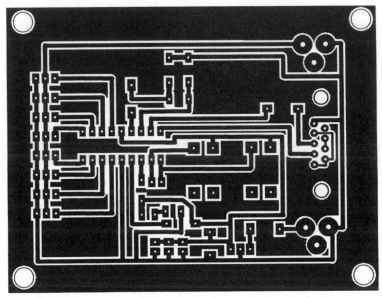

■ **FIGURE 5.58** *PCB foil pattern for the serial servo controller circuit board.*

Circuit Board Drilling and Parts Placement

Use a 1/32-inch drill bit to drill all of the component holes on the PCB. Drill the holes for the voltage regulator (U1) and diode (D1) with a 3/64-inch drill bit. Use **Table 5.2** and **Figure 5.59** to place the parts on the component side of the circuit board. The PIC 16F84 microcontroller (U2) is mounted in an 18-pin IC socket. The 18-pin socket is soldered to the PC board, and the PIC will be inserted after it has been programmed. Use a fine-toothed saw to cut the board along the guide lines, and drill the mounting holes on the corners using a 5/32-inch drill bit. Use four 1 1/4-inch standoffs to mount the board. **Figure 5.60** shows the finished main controller board.

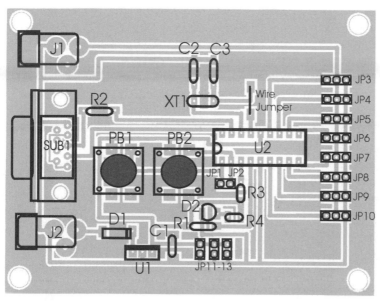

■ **FIGURE 5.59**   *Parts placement diagram for serial servo controller circuit board.*

■ **FIGURE 5.60**   *Finished serial servo controller circuit board.*

Programming the PIC 16F84A

The PIC 16F84A software for the serial servo controller operates by receiving information in a serial format from a host computer. The PicBasic Pro compiler, by microEngineering Labs (www.melabs.com), includes an instruction called *serin* that makes receiving serial information easy. When using the *serin* command to receive data, PicBasic Pro lets you define a qualifier enclosed within brackets before any more data are received. *Serin* must receive the qualifier bytes in exact order before receiving any data items. If any of the bytes received do not match the bytes in the qualifier sequence, the qualification process starts over (i.e., the next received byte is compared to the first item in the qualifier list). Another feature of the *serin* command is the ability to set a *timeout* and *label*. The *timeout* qualifier allows the program to continue execution at the point specified by *label* if a character is not received within a certain amount of time. *Timeout* is specified in 1-ms units. This is ideal for our application because it gives our program the ability to constantly update the servos with positional information if nothing is happening on the communications port.

For our program we will use the ASCII value of 255 as the qualifier to identify that data are being sent from the host computer. Once the ASCII value of 255 has been received, the next byte of information will be stored in a variable called *slider*. This variable determines which servo will be updated. The next byte of information is stored in a variable called *control* and contains the position to which the servo will be set. The line of code is shown below.

serin com_in,baud,5,set_pos,[255],slider,control

The serial data are received on pin *com_in* (PORTA.0) at a baud rate of *baud* (2400). If nothing is received on pin 0 of port A within 5 ms, then program execution will continue at *set_pos*, where the servo positions are set. When the serial data that determine which servo and its position are received, the PIC will then transmit the data back to the computer on pin *com_out* (PORTA.1). The serial servo control program running on the computer will use this information to verify that the PIC has received the correct information and will then send the next set of servo and position information. The program execution will remain in a tight loop of receiving servo position data and then setting the servos with the data received.

The serial servo controller program is called ssc.bas and is listed in **Program 5.1**. Compile the program and then program the PIC 16F84A with the ssc.hex file listed in **Program 5.2**. Place the programmed PIC 16F84A in the 18-pin socket on the robot's main board with the notch located toward the two push-button switches, as shown in **Figure 5.59**. Now that the hardware section of the robot arm is finished, it is time to connect all of the parts together.

■ **PROGRAM 5.1** *ssc.bas program listing.*

```
TRISB = %00000000
TRISA = %00001101

DEFINE OSC 4

include "modedefs.bas"

Led          VAR PORTA.4
Baud         CON N2400
Com_In       VAR PORTA.0
Com_Out      VAR PORTA.1
Control      VAR BYTE
Slider       VAR BYTE
Sync         VAR BYTE
I            VAR BYTE
S0           VAR BYTE
S1           VAR BYTE
S2           VAR BYTE
S3           VAR BYTE
S4           VAR BYTE
S5           VAR BYTE
S6           VAR BYTE
S7           VAR BYTE

LOW PORTB.0
LOW PORTB.1
LOW PORTB.2
LOW PORTB.3
LOW PORTB.4
LOW PORTB.5
LOW PORTB.6
LOW PORTB.7
HIGH Led

S0 = 127
S1 = 111
S2 = 144
S3 = 65
S4 = 127
S5 = 127
S6 = 127
```

```
        S7 = 127

    Start:

          SERIN Com_In,Baud,7,Set_Pos,[255],Slider,Control
          SEROUT  Com_Out,Baud,[Slider,Control]

          IF Slider = 0 THEN S0 = Control

          IF Slider = 1 THEN
             S1 = Control
             S2 = 254 - S1
          ENDIF

          IF Slider = 2 THEN S3 = Control

          IF Slider = 3 THEN
             S4 = Control
             S5 = 254 - S4
          ENDIF

          IF Slider = 4 THEN S6 = Control

          IF Slider = 5 THEN S7 = Control

    Set_Pos:

          PULSOUT PORTB.0,S0
          PULSOUT PORTB.1,S1
          PULSOUT PORTB.2,S2
          PULSOUT PORTB.3,S3
          PULSOUT PORTB.4,S4
          PULSOUT PORTB.5,S5
          PULSOUT PORTB.6,S6
          PULSOUT PORTB.7,S7

    GOTO Start
```

■ **PROGRAM 5.2**   *ssc.hex file listing.*

```
:10000000B428A00008820000C080D040319AF28A920EB
:100010008413200088000664000D280E288C0A03191A
:100020008D0F0B288006AF2823088C0021088D0037
```

:1000300001308E008F0164003C20031C2F288E0BA2
:100040001B28FF308F0703181B288C07031C8D0704
:10005000031CAF2832308E0000308F001B286D202B
:1000600008308F006E203C208E0C36288F0B3228F3
:100070006E2003140E0808002208840020088417 4C
:1000800080004841300051F192006FF3E080092001B
:100090002208840009309300031053209 20C930B24
:1000A0004D280314532884139F1D62280008200440
:1000B0001F1D2006800084170008200 4031C200652
:1000C00080006E2800082004031C20061F1920064B
:1000D00080008417200980056E281F171F0D063920
:1000E0008C007C208D008C0A7C201F1F8D281F1304
:1000F0008C000230A4208D2800308A000C08820772
:100100000013475340334153400343C340C34D934A0
:10011000FF3A84178005AF288D01E83E8C008D09D9
:10012000FC30031C96288C07031893288C07640066
:100130008D0F93280C189C288C1CA0280000A02848
:10014000080003108D0C8C0CFF3E0318A1280C082E
:10015000AF288C098D098C0A03198D0A08008313B6
:1001600003138312640008008316 86010D30850096
:10017000831206108316061083128610 83168610CB
:10018000831206118316061183128611 83168611B7
:10019000831206128316061283128612 83168612A3
:1001A000831206138316061383128613 8316 8613 8F
:1001B00083120516831605128 3127F30A6006F3056
:1001C000A7009030A8004130A9007F30AA007F30FE
:1001D000AB007F30AC007F30AD000530A2000130B5
:1001E000A00004309F000730A300A1011420031CCD
:1001F0003E29FF3C031DF6281420031C3E29AE00B7
:1002000001420031C3E29A4000530A2000230A000E7
:1002100004309F002E0847202408472064002E0841
:1002200000 03C031D15292408A60064002E08013C8B
:1002300000 31D1F292408A7002708FE3CA80064000E
:100240002E08023C031D26292408A90064002E085C
:1002500033C031D30292408AA002A08FE3CAB00F9
:1002600064002E08043C031D37292408AC006400F8
:100270002E08053C031D3E292408AD0026088C00ED
:100280008D0106308400013001202708 8C008D018B
:100290000063084000230012028088C008D01063 0D1
:1002A0008400043001202908 8C008D010630840070
:1002B0000083001202A088C008D0106308400103 09F
:1002C00001202B088C008D0106308400203001209 5
:1002D0002C088C008D010630840040300120 2D0850

:0E02E0008C008D010630840080300120ED2856
:02400E00F53F7C
:00000001FF

Use **Figures 5.61** and **5.62** to connect each of the servos to the serial servo controller board. Note that the control wire of each servo is plugged into the servo connectors with the yellow or white wire situated closest to the PIC 16F84A, and the black wire is closest to the edge of the board. Plug one end of a straight-through serial cable into the controller board, and the other end into the serial port on a PC. When the Visual Basic control software is completed in the next chapter and the arm is ready to be tested, plug the 6-V DC and 9-V DC adapters into the serial servo controller board, as indicated in **Figures 5.61** and **5.62**. The completed hardware for the robot arm system is shown in **Figure 5.63**.

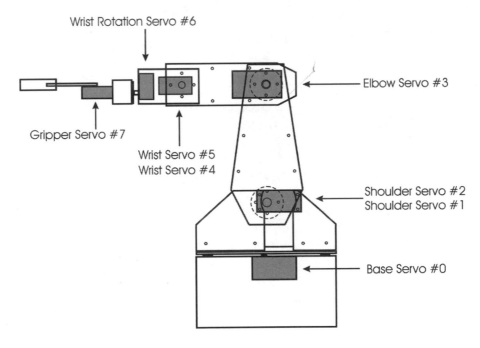

■ **FIGURE 5.61**   *Servo identification diagram.*

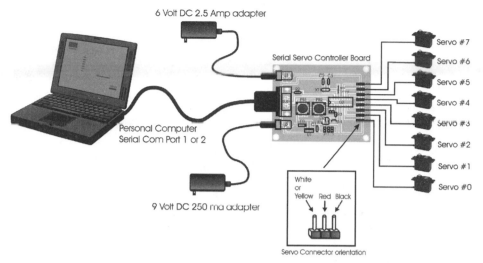

■ **FIGURE 5.62** *Electronics connection diagram.*

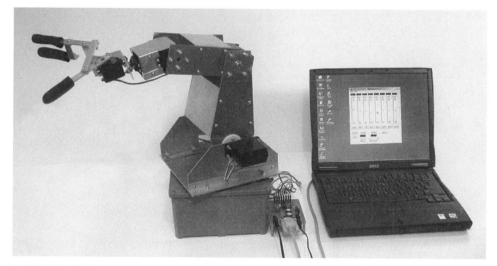

■ **FIGURE 5.63** *Completed hardware of the robot arm system.*

Summary

Now that the mechanics and electronics communications interface of the robot arm project are complete, a Visual Basic 6 software application will be developed so that the arm can be controlled directly from a PC. A robot arm is an essential component to a functional humanoid robot because it allows the robot to grasp and carry objects so that it can do useful work.

Visual Basic 6 Robot Interface Software

This chapter will focus on designing a serial interface application that will be used to control the serial servo controller circuit board that was built in Chapter 5 to run the robot arm. The serial servo interface software will be built using Microsoft's Visual Basic 6. The ability to design your own serial interface application that runs on any personal computer (PC) has never been easier using Visual Basic 6. Using a microcontroller in conjunction with a PC provides enormous power and flexibility to any project. The following project can be used as shown, or the concepts can be adapted and applied to any number of projects that require serial communications software development.

The serial servo control software will allow you to control up to eight servos with an easy-to-use graphics interface. Moving the position of the sliders on the graphics interface controls each of the eight servos. When the robot arm is positioned where desired, the eight servo settings can be saved in a control sequence with the click of a button. When you have recorded all of the arm positions into a sequence, the sequence of stored positions can be played back at varying movement speeds. The serial servo controller application is shown in **Figure 6.1**. You can download this project at www.thinkbotics.com to save you from typing in the code. Although downloading the code is a lot easier, you can learn a lot about Visual Basic programming by working your way through the project. The compiled executable and install package can also be downloaded from www.thinkbotics.com for free.

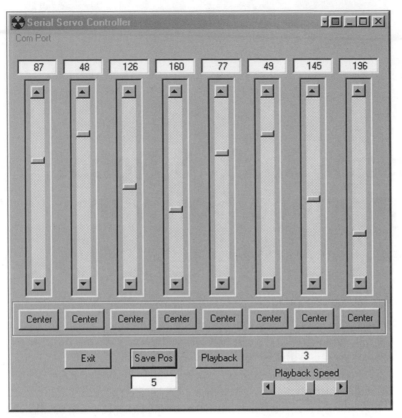

■ **FIGURE 6.1** *The serial servo controller application.*

Creating the Serial Servo Controller Project

When you start Visual Basic 6.0 from the Start Programs menu, you should see a screen similar to the one shown in **Figure 6.2**. Choose the standard EXE project, and then click the Open button in the lower right-hand corner.

Because you have instructed Visual Basic to create a standard Windows application, the next screen you see should look similar to **Figure 6.3**. An empty form has been provided, along with all the tools necessary to populate the form with controls. The tools that are presently available are contained in the toolbox. The toolbar may be situated horizontally at the top of the window or at the side. The form can be resized by clicking on the flat gray area to select it, and then by clicking and holding the mouse button over one of the squares while dragging the form to resize it. You will need to resize the form occasionally as controls are added. This form will eventually turn into our completed serial servo controller application and will look like the screen capture in **Figure 6.1**.

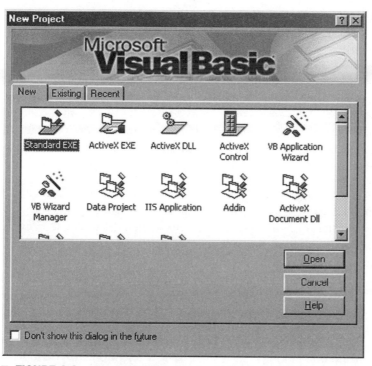

■ **FIGURE 6.2** *Start with the standard EXE project type.*

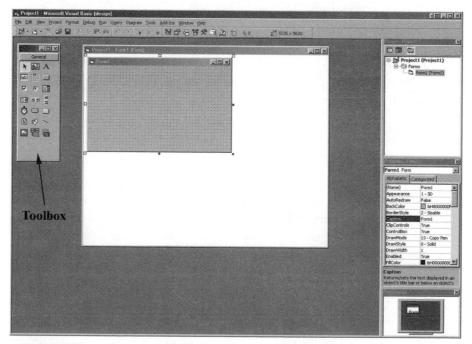

Toolbox

■ **FIGURE 6.3** *The Visual Basic development environment.*

Adding the Microsoft Comm 6.0 Control

To the left of the Form window is the toolbox. The toolbox contains buttons or icons representing each of the types of controls that are commonly added to forms. There are many more controls that can be added to the toolbox. To communicate with the robot arm via the serial servo controller circuit board, we will need to add the MSComm 6.0 control. The MSComm control was one of the first custom controls designed for Visual Basic and is the original means of communicating through the serial port. The MSComm control provides serial communications for your application by allowing the transmission and reception of data through the serial ports.

The MSComm control provides two ways to handle communications. The first is event-driven communications, a very powerful method for handling serial port interactions. In many situations, you may want to be notified the moment an event takes place, such as when a character arrives or a change occurs in the Carrier Detect (CD) or Request To Send (RTS) lines. In such cases, you can use the MSComm control's OnComm event to trap and handle these communications events. The second method of handling communications is by polling for events and errors by checking the value of the CommEvent property after each critical function of your program.

The MSComm control 6.0 icon looks like a small yellow telephone. If you look at the toolbox, you will notice that it is missing at this point. To add the MSComm control to the toolbox, follow the steps below.

1. On the menu bar at the top of the screen, click on Project.
2. Scroll down and click on Components.
3. On the next screen that opens up, scroll down until you see the Microsoft Comm Control 6.0.
4. Place a check mark in the box to the left of the Microsoft Comm Control by clicking on the box, as shown in **Figure 6.4**.
5. Click Apply at the bottom of the screen, and then click on the Close button.

You should now see the MSComm control icon on the toolbar. **Figure 6.5** identifies the Comm control, along with some of the other controls that we will be using to build the application.

Adding Controls to the Form

On the toolbar, place the mouse over the Comm control and click to select it. Next, place the mouse over the lower left area on the blank form. Click and hold the

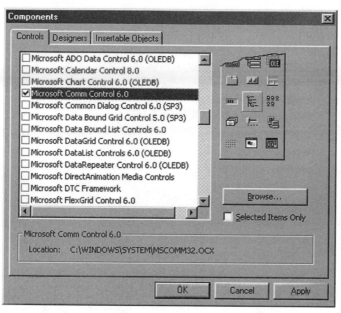

■ **FIGURE 6.4** *Adding the MSComm Control 6.0.*

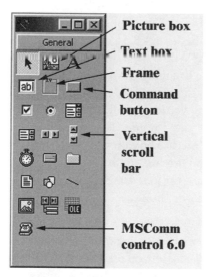

Picture box
Text box
Frame
Command button
Vertical scroll bar
MSComm control 6.0

■ **FIGURE 6.5** *Toolbox with the MSComm control added.*

mouse button down while dragging down and to the right. When you release the mouse button, a yellow telephone should appear in the lower left corner of the form, like one shown in **Figure 6.6**.

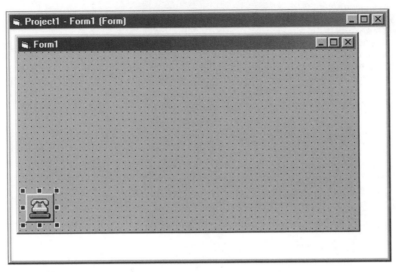

■ **FIGURE 6.6**   *MSComm control added to the form.*

Next, we will add eight scroll bars to the blank form. The scroll bars will be used to control the position of the servo motors. On the Visual Basic toolbox, click on the vertical scroll bar to select it. Move the mouse to the top left side of the form, then click and hold the mouse button down while dragging down and to the right. You should now see a single vertical scroll bar on the left side of the form. You can now resize the scroll bar to any size. Keep placing vertical scroll bars on the form until you have a total of eight vertical scroll bars side by side. To give each of the sliders a three-dimensional look, you can draw a picture box around each slider. If you added picture boxes, move the mouse over the first picture box, click the right mouse button, and then choose the Send to Back option. Do this for all of the remaining picture boxes. When finished, your form should look very similar to **Figure 6.7**.

The next controls to be added to the form will be eight command buttons that will be used to center each of the servos when clicked. The button associated with each servo will send data to the Serial Servo Controller to center the servo, move the scroll bar position indicator to its middle position, and then change the value displayed in the text box above the scroll bar.

To add command buttons to the form, click the command button on the toolbox (see **Figure 6.5**), and then place one command button below each scroll bar on the form by clicking and dragging the mouse where you would like to place the but-

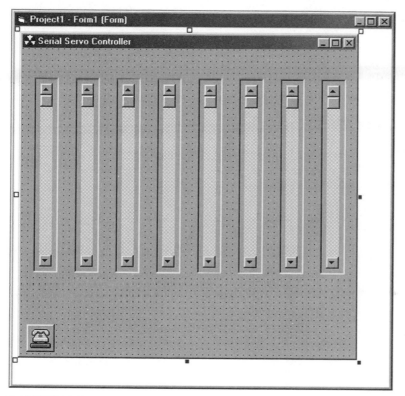

■ **FIGURE 6.7**   *Vertical sliders added to the form.*

ton. When you are finished, there should be eight command buttons situated below the vertical scroll bars. Follow the steps below to change the text on each command button.

1. Click on the first command button on the left side of the form underneath the first vertical scroll bar.

2. On the right side of the Visual Basic screen, you will see a Properties menu like the one shown in **Figure 6.8**.

3. Click on the area inside the Properties box marked Caption. Click on the area just to the right of the highlighted area. Delete whatever is in this box (probably the word Command1) and type the word Center into this area. You should now see the word Center displayed in the middle of the first command button. Follow the same procedure to change the text in each command button to the word Center. You can add a frame around the command buttons by clicking on Frame in the toolbox, and then drawing a frame around the buttons on the form. In the Properties box for the frame, delete the caption so that nothing

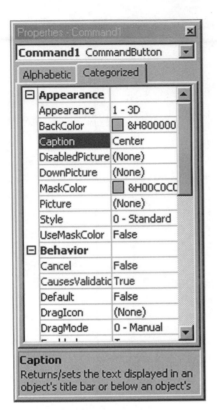

■ **FIGURE 6.8**   *Control Properties box.*

but the frame is displayed. Right-click the mouse button on the frame, and then choose the Send to Back option. Click the main form to select it. To change the name of the main form, move to the properties box and select the caption. Change the text in the caption box to Serial Servo Controller or whatever name you would like. Scroll down through the properties box and select Icon. Click on the button with three dots to open the selection window. Explore to: C:\Program Files\ Microsoft Visual Studio\Common\graphics\Icons\Misc\Misc01.ico or any other icon that you like. Click the Open button. The title and icon that you selected should be displayed at the top of the form. The form with command buttons, a title, and an icon added is shown in **Figure 6.9**.

The next step will be to add some text boxes to the form. The text boxes will display the actual values of the scroll bars, and will contain the data that will be sent out of the serial port to the serial servo controller board to which the robot arm is connected. Click on the text box in the toolbox, and then place a text box over the first slider. To handle the data in the text boxes, it will be to our advantage to cre-

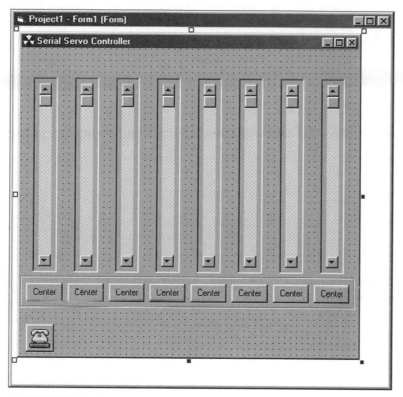

■ **FIGURE 6.9** *Command buttons, title, and icon added to the form.*

ate a control array of text boxes so that they can be indexed easily when data needs to be updated or sent to the serial port.

To create a second text box that is part of a control array, click on the first text box, then right-click and select copy. Click on the main form, right-click, and then select Paste. You will see a dialog like the one shown in **Figure 6.10** asking if you would like to create a control array. Select Yes. Position the text box that was just pasted onto the form over the second slider. You will notice in the Properties window that this text box is labeled Text1(1). If you click on the first text box, the Properties window will show that it is labeled Text1(0). Keep copying text boxes and placing them over the vertical scroll bars until there are eight in total. The last text box should have been automatically named Text1(7). Being able to address the text boxes as an array will simplify the code when we add the ability to record and play back the various servo positions.

Now that all of the text boxes are in position, select the Text1(0) text box on the far left. Select the text property in the Properties box and delete all of the characters until this area is blank. We don't actually want anything in the text boxes until the code for the vertical scroll bars sends data there. Select each of the remaining

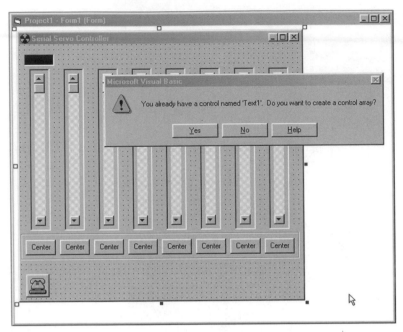

■ **FIGURE 6.10**   *Setting up a control array of text boxes.*

text boxes and delete the text property of each box so that they are all blank. When you are finished, each text box above the sliders should be blank. You can change the look of the text boxes by changing the BackColor and ForeColor setting in the Properties box. The ForeColor setting changes the color of the text and the BackColor setting changes the color of the area behind the text. Try experimenting with these settings until you find a color scheme that you like. You should now have a project that looks like **Figure 6.11**.

Adding a Menu

The project will use a menu located at the top of the form to allow the user to select between serial port 1 or port 2. To add a menu to the Visual Basic project, follow the steps below:

1. Click on the main form to select it.
2. Select Tools from the top menu.
3. Scroll down to Menu Editor and click on it.

You should see a form similar to the one shown in **Figure 6.12**. In the first box titled Caption, type Com Port. The caption will be displayed as the menu at the top of the form. In the next box titled Name, type mnuSelectPort exactly as shown in **Figure 6.12**. Click the button marked Next just above the white window on the

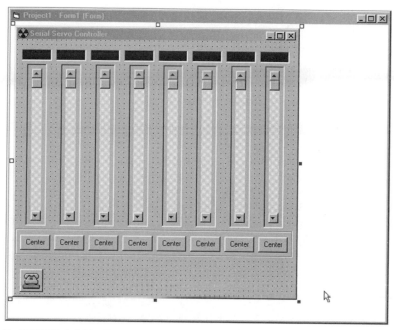

■ **FIGURE 6.11**   *Text boxes added to the form.*

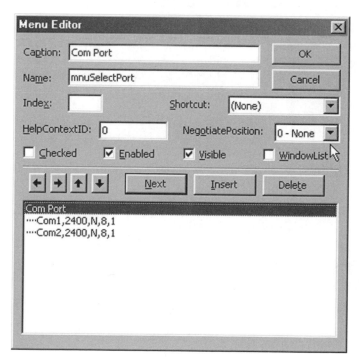

■ **FIGURE 6.12**   *Menu editor screen.*

bottom of the menu editor screen. Type in Com1,2400,N,8,1 beside Caption. The text that you are typing will be displayed in the large white box just below where it says Com Port, highlighted in blue. If you don't see four periods before this entry that look like ....Com1,2400,N,8,1 then click the arrow pointing to the right. Your entry should look exactly like the one in **Figure 6.12**. The periods make the other menu items submenus inside the main menu selection.

In the box marked Name, type: mnuCom 12400. Repeat this same procedure for the next menu item. Click next and type Com2,2400,N,8,1 next to Caption. Below this, next to Name, type mnuCom22400. You should now have a screen that looks exactly like **Figure 6.12**.

The next step is to add the controls that will give our application the ability to save sequences of arm positions, and then play them back at various speeds. We will also add an exit button. Use **Figure 6.13** as a guide to place three more command buttons. Change the caption for the first button to Exit. The second button cap-

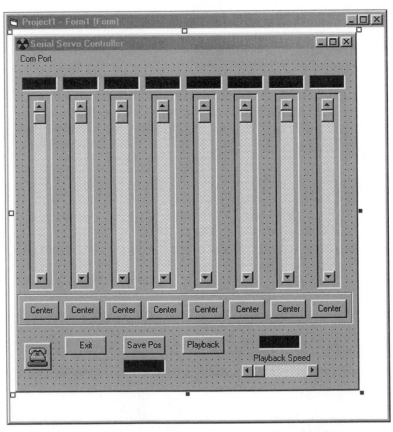

■ **FIGURE 6.13**   *Form with record and playback controls added.*

tion is Save Pos, and the caption for the third button is Playback. Add a text box underneath the Save Pos button and another text box beside the Playback button. Place a horizontal scroll bar underneath the last text box that was placed on the form. Draw a label box above the horizontal scroll bar and change the Caption to Playback Speed. When all of the controls have been added, the form should look very similar to **Figure 6.13**.

Now that all of the controls that make up the serial servo controller application have been placed on the form, it is time to assign the code that goes with each one. This is accomplished by coding instructions into each control. Because Visual Basic is event-driven, the coded instructions are activated only when the controls are used. This means that code will not be executed if none of the controls in our project are being used.

The Form_Load Event

When a Visual Basic application is first started and the main form is loaded, it is most often desirable to execute configuration code before anything else happens. The area that is reserved for these kinds of instructions is called Form_Load. For this application, we will tell the form to run code that will perform the following functions.

1. Set up the Communications Port with the operating parameters of 2400,N,8,1.
2. Select whether to use Com port 1 or Com port 2.
3. Open the Com port to make it available for use.

To enter the code for the Form_Load event, double-click on a blank spot on the main form. A window will open up similar to the one shown in **Figure 6.14**, but it will be blank between the Private Sub Form_Load( ) and End Sub. Type in everything between the Private Sub Form_Load( ) and End Sub exactly as shown below and in **Figure 6.14**. Note that **Figure 6.14** includes comments that can be left out.

```
Private Sub Form_Load()
    MSComm1.Settings = "2400,N,8,1"
    MSComm1.CommPort = 1
    MSComm1.PortOpen = True
    pos_index = 1
    speed = 1
    Text2.Text = pos_index
    Text3.Text = speed
    Dim I As Byte

    For I = 0 To 7
```

```
        Text1(I).Text = Str(127)
        MSComm1.Output = Chr$(255)
        MSComm1.Output = Chr$(I)
        MSComm1.Output = Chr$(127)
    Next I

    VScroll1.Value = Text1(0).Text
    VScroll2.Value = Text1(1).Text
    VScroll3.Value = Text1(2).Text
    VScroll4.Value = Text1(3).Text
    VScroll5.Value = Text1(4).Text
    VScroll6.Value = Text1(5).Text
    VScroll7.Value = Text1(6).Text
    VScroll8.Value = Text1(7).Text
End Sub
```

When you double-clicked on the form, the code view screen came up with the Private Sub Form_Load( ) and End Sub already typed for you. This will be the case each time you double-click on any control item you place on your forms. The reason for this is to identify the code area that corresponds to the control that was selected.

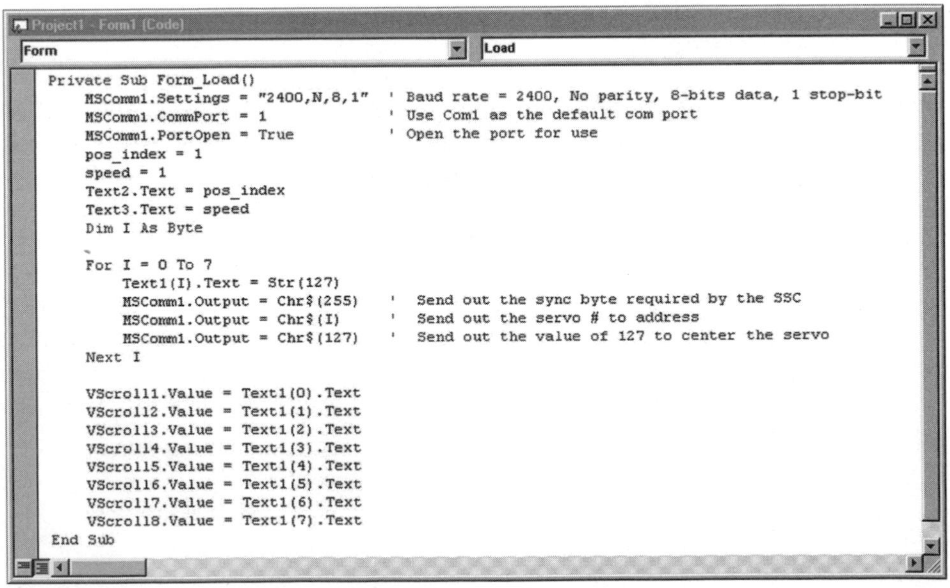

■ **FIGURE 6.14** *Code window view for the Form_Load event.*

Scroll Bars

Now that the default communications parameters have been set up using the Form_Load event, it is time to make the scroll bars do something. The scroll bars will actually perform a couple of functions for this project. First, the scroll bars provide an interface to the user that can be moved with the mouse. Servo position data will be created and sent to the serial servo controller, depending on the position of the slider. Each slider will also send a unique number to the serial servo controller, identifying which servo will be updated with the position data.

The actual value of each scroll bar will be displayed in the text boxes above each slider to provide the user with a visual representation of the data that are being sent out of the serial port to the robot arm serial servo controller. Each time a scroll bar is moved, the text box directly above it will display the actual servo position data that are currently being sent through the serial port to the serial servo controller. The scroll bar code will also tell the serial servo controller which servo should be moved. The first byte sent is a value of 255, which is used by the microcontroller code of the serial servo controller circuit as a qualifier or sync byte, to let it know that data are being sent.

Place your mouse over the first scroll bar on the left side of the project form and click on it once to select it. In the Properties menu on the right side of the project window, you should see the properties settings called Max and Min, as shown in **Figure 6.15**. This area lets us set the Maximum and Minimum values for the scroll bar. The Maximum and Minimum values that we want to send to the serial servo controller for the position data (byte 3) are from 1 to 254. This is the range of information that we want the scroll bar to produce when it is moved all the way up (Min), or all the way down (Max). In the properties window for the first scroll bar, assign the Max value for the first scroll bar a value of 254, and make the Min value 1. If you want to swap these values, the movement of the scroll bars will reverse the operation of the servo motors when the scroll bars are moved. Before moving to the next step, assign these same values to all of the remaining seven scroll bars.

Select the first scroll bar again and look in the Properties box. Find the box for LargeChange and SmallChange and change the values of each one to a 1 if they are not already set to 1. This will give very precise control over each servo motor when you click and hold your mouse on the up or down arrows of each scroll bar. Set each of the remaining scroll bars LargeChange and SmallChange values in the Properties box to a 1.

Scroll Bar Code

Double-click on the first scroll bar on the left side of the project form to bring up the code window. It will look something like the one shown in **Figure 6.16**, but yours will be blank between Private Sub VScroll1_Change( ) and End Sub.

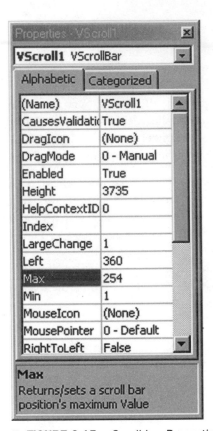

Properties - VScroll1

VScroll1 VScrollBar

Alphabetic | Categorized

(Name)	VScroll1
CausesValidatic	True
DragIcon	(None)
DragMode	0 - Manual
Enabled	True
Height	3735
HelpContextID	0
Index	
LargeChange	1
Left	360
Max	254
Min	1
MouseIcon	(None)
MousePointer	0 - Default
RightToLeft	False

Max
Returns/sets a scroll bar
position's maximum Value

■ **FIGURE 6.15** *Scroll bar Properties window.*

The subroutine for the first scroll bar is named VScroll1_Change() because when the scroll bar is moved up or down, changes are occurring. The code in this routine is activated each time that there is a change made to the scroll bar's position. The code for the first scroll bar is shown below:

```
Private Sub VScroll1_Change()
    Text1(0).Text = Str(VScroll1.Value)
    MSComm1.Output = Chr$(255)
    MSComm1.Output = Chr$(0)
    MSComm1.Output = Chr$(Text1(0).Text)
End Sub
```

Note that the scroll bar subroutine affects only the text box above it. In this case, the routine Private Sub VScroll1_Change() affects only Text1(0).Text. All of the text boxes are part of an array, and each text box is indexed by the number in the brackets. The first text box is Text1(0).Text, the second is Text1(1).Text, the third is Text1(2).Text, etc. The decision to use an array for the text box was made so

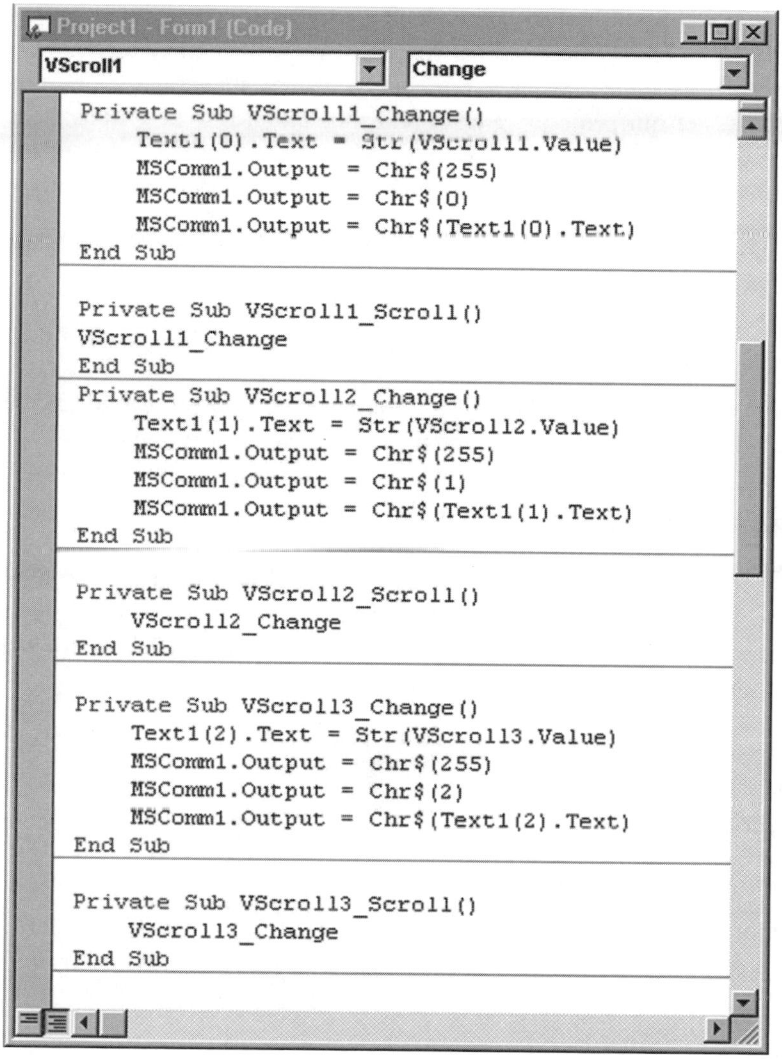

```
Project1 - Form1 (Code)                              _ □ X

VScroll1                      ▼   Change                    ▼

    Private Sub VScroll1_Change()
        Text1(0).Text = Str(VScroll1.Value)
        MSComm1.Output = Chr$(255)
        MSComm1.Output = Chr$(0)
        MSComm1.Output = Chr$(Text1(0).Text)
    End Sub

    Private Sub VScroll1_Scroll()
    VScroll1_Change
    End Sub
    Private Sub VScroll2_Change()
        Text1(1).Text = Str(VScroll2.Value)
        MSComm1.Output = Chr$(255)
        MSComm1.Output = Chr$(1)
        MSComm1.Output = Chr$(Text1(1).Text)
    End Sub

    Private Sub VScroll2_Scroll()
        VScroll2_Change
    End Sub

    Private Sub VScroll3_Change()
        Text1(2).Text = Str(VScroll3.Value)
        MSComm1.Output = Chr$(255)
        MSComm1.Output = Chr$(2)
        MSComm1.Output = Chr$(Text1(2).Text)
    End Sub

    Private Sub VScroll3_Scroll()
        VScroll3_Change
    End Sub
```

■ **FIGURE 6.16** *Code view window showing scroll bar code.*

that storing the servo positions would be simplified when we add the code to record and play back the arm movements. Also note that scroll bar 1 sends out the hard-coded value of Chr$(0). This is because scroll bar 1 sends information to the robot arm to control servo number 0. Each scroll bar sends out the sync byte to tell the serial servo controller that information is being sent, the number of the servo that will be moved, and the position stored in Text1(0 to 7).Text. Add the listed code above for each scroll bar, being sure to change the Text1(x).Text index number, Vscroll number, and Servo number to the text box number that is above the scroll bar for which you are entering the code. See **Figure 6.16** to get the idea.

A scroll bar will normally only display its changed value once you let go of the mouse button. Without telling the scroll bar to update its value immediately, it will wait until the scroll bar slider control is released to display its actual value in its associated text box. For our project, we want real-time updates to be visible in the text boxes, and we also want real-time information to be sent immediately to the serial port. The code below causes the text box contents to be updated immediately when the scroll bar is being dragged.

```
Private Sub VScroll1_Scroll()
VScroll1_Change
End Sub
```

Add this code for each of the scroll bars. Try commenting out the code and see the effect it has on how the individual servos of the robot arm are controlled.

Command Button Code

The next step is to add some basic code to our command buttons. The first eight of our command buttons will do nothing but send the command to the serial servo controller to move the servo to its center position. The code view window for the first three command buttons is shown in **Figure 6.17**.

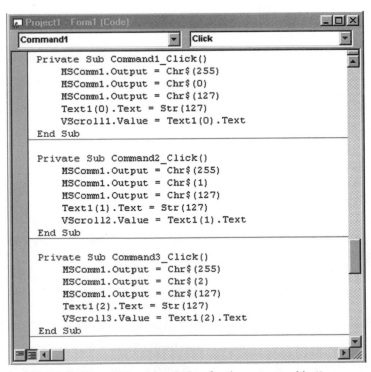

```
Project1 - Form1 (Code)                                    _ □ ×
Command1                    ▼    Click                       ▼
    Private Sub Command1_Click()
        MSComm1.Output = Chr$(255)
        MSComm1.Output = Chr$(0)
        MSComm1.Output = Chr$(127)
        Text1(0).Text = Str(127)
        VScroll1.Value = Text1(0).Text
    End Sub

    Private Sub Command2_Click()
        MSComm1.Output = Chr$(255)
        MSComm1.Output = Chr$(1)
        MSComm1.Output = Chr$(127)
        Text1(1).Text = Str(127)
        VScroll2.Value = Text1(1).Text
    End Sub

    Private Sub Command3_Click()
        MSComm1.Output = Chr$(255)
        MSComm1.Output = Chr$(2)
        MSComm1.Output = Chr$(127)
        Text1(2).Text = Str(127)
        VScroll3.Value = Text1(2).Text
    End Sub
```

■ **FIGURE 6.17**   *Code view window for the command buttons.*

Begin entering code for each command button by double-clicking on the first command button on the left side of the project window. The code window will be displayed. Enter the code listed below in the area between the Private Sub Command1_Click() and End Sub. Repeat the same procedure for each of the Center buttons. Be sure to change the servo number and the text box index number to the proper values for each button.

```
Private Sub Command1_Click()
    MSComm1.Output = Chr$(255)
    MSComm1.Output = Chr$(0)
    MSComm1.Output = Chr$(127)
    Text1(0).Text = Str(127)
    VScroll1.Value = Text1(0).Text
End Sub
```

The next command button for which we will add code will be the Exit button. When exiting any serial communications program, it is important that the open port be closed first. We will check to see if the port is open first and close it only if it is open. Double-click the Exit button to bring up the code window. Enter the code below in the space between Private Sub Command9_Click() and End Sub.

```
Private Sub Command9_Click()
    If MSComm1.PortOpen = True Then
        MSComm1.PortOpen = False
    End If
    End
End Sub
```

Menu Options

The next step is to add the code to the menu options for changing the Com port. The menu selections will offer the user the option of changing the active Com port. To start entering the code for the menu selections, click on the main menu at the top of the form marked Com Port. Scroll down to the menu selection marked Com1 2400,N,8,1, click on it, and the code window will pop up. Enter the following code between the Private Sub mnuCom12400_Click( ) and the End Sub statements.

```
Private Sub mnuCom12400_Click()
    If MSComm1.PortOpen = True Then
        MSComm1.PortOpen = False
    End If
    MSComm1.Settings = "2400,N,8,1"
    MSComm1.CommPort = 1
```

```
    MSComm1.PortOpen = True
End Sub
```

Follow the same procedure for the next menu selection marked Com2 2400,N,8,1 and enter the following code between the Private Sub mnuCom22400_Click( ) and the End Sub statements.

```
Private Sub mnuCom22400_Click()
    On Error GoTo Error
    If MSComm1.PortOpen = True Then
        MSComm1.PortOpen = False
    End If
    MSComm1.Settings = "2400,N,8,1"
    MSComm1.CommPort = 2
    MSComm1.PortOpen = True
    GoTo No_error
Error:
    MSComm1.Settings = "2400,N,8,1"
    MSComm1.CommPort = 1
    MSComm1.PortOpen = True
    MsgBox ("Invalid port - will use Port 1 instead")
No_error:
End Sub
```

The first line of code for Com port 2 tells the program to resume execution at the label called Error if an error is encountered. The reason for including this code is because many computers (like the one I have at home) only possess one Com port. If the second Com port does not exist, then an error will be generated and the program will crash. If Com Port 2 does not exist, then it is set back to Com Port 1 and the user is notified with a message box. There are other ways to check to see if the port actually exists before setting it, but for our application, this will suffice. **Figure 6.18** shows the code view window for the menu settings.

Record and Playback

All that remains to complete the application is to add code for the record and playback section. Double-click the Save Pos button to bring up the code window. Enter the code listed below in the space between Private Sub Command10_Click() and End Sub.

```
Private Sub Command10_Click()
    Dim I As Byte
    I = 0
    ReDim Preserve stored_positions(0 To 7, 1 To pos_index)
```

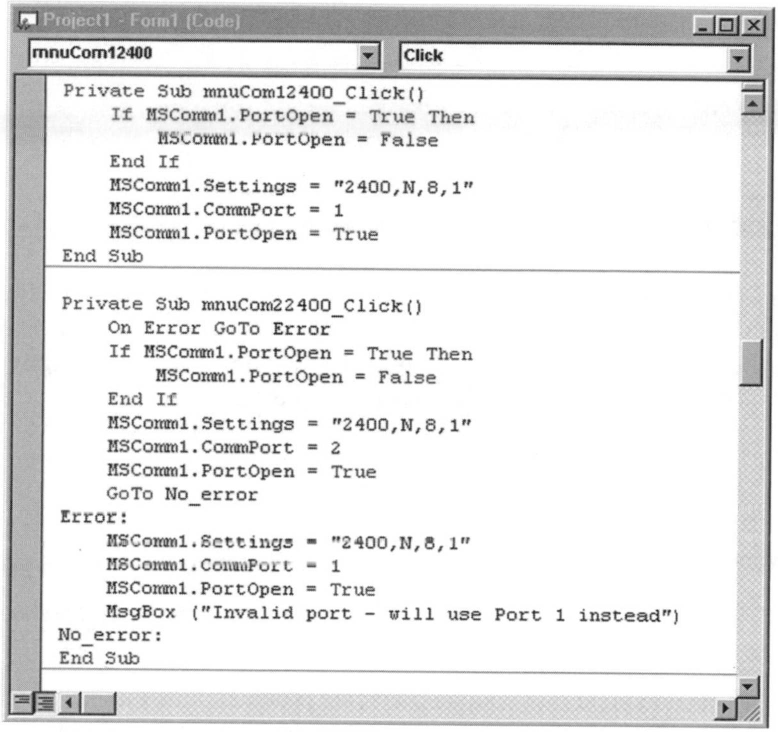

■ **FIGURE 6.18** *Code view window for the menu settings.*

```
For I = 0 To 7
    stored_positions(I, pos_index) = Str(Text1(I).Text)
Next I
pos_index = pos_index + 1
Text2.Text = pos_index
End Sub
```

When the Store Pos button is clicked, all of the current servo position values are stored in the stored_positions array. The array named stored_positions needs to be defined in the General Declarations at the very top of the code window. Open the code window and select General Declarations from the dropdown menu. Enter the following declarations into that area.

```
Dim stored_positions() As Double
Dim pos_index As Double
Dim speed As Byte
Dim InString As String
```

The code window with the General Declarations selected is shown in **Figure 6.19**.

```
Project1 - Form1 (Code)                                    _ □ ×
(General)                      ▼   (Declarations)              ▼
    Dim stored_positions() As Double
    Dim pos_index As Double
    Dim speed As Byte
    Dim InString As String
```

■ **FIGURE 6.19** *Code view window showing General Declarations.*

Click on the horizontal scroll bar underneath the Playback Speed Label. In the properties box for the scroll bar, change the Max value to 5 and the Min value to 1. Double-click on the scroll bar to open the code view window. Enter the following code between Private Sub HScroll1_Change() and End Sub.

```
Private Sub HScroll1_Change()
    Text3.Text = Str(HScroll1.Value)
    speed = HScroll1.Value
End Sub
```

The contents of the variable *speed* will be used as the delay value in seconds when the Playback button is activated.

Double-click on the Playback button to open the code view window. Enter the following code in the area between Private Sub Command11_Click() and End Sub.

```
Private Sub Command11_Click()
Dim I As Byte
Dim J As Byte
Dim K As Byte
Dim slider As Byte
Dim control As Byte
Dim start_time As Double
Dim end_time As Double
I = 0
J = 0
K = 1
slider = 9
control = 0

If pos_index = 1 Then GoTo Finish

  Do While K < pos_index
    Text2.Text = K
    Text2.Refresh
    For I = 0 To 7
```

```
            Text1(I).Text = stored_positions(I, K)
            Text1(I).Refresh
            DoEvents
            J = 0
            Do While (slider <> I) And (control <> Asc(Text1(I).Text))
                MSComm1.Output = Chr$(255)
                MSComm1.Output = Chr$(I)
                MSComm1.Output = Chr$(Text1(I).Text)

                MSComm1.InputLen = 0
                If MSComm1.InBufferCount > 1 Then
                    InString = MSComm1.Input
                    slider = Asc(Mid(InString, 1, 1))
                    control = Asc(Mid(InString, 2, 1))
                End If
            J = J + 1
            If J = 50 Then GoTo Com_error
            Loop
        Next I

            VScroll1.Value = Text1(0).Text
            VScroll2.Value = Text1(1).Text
            VScroll3.Value = Text1(2).Text
            VScroll4.Value = Text1(3).Text
            VScroll5.Value = Text1(4).Text
            VScroll6.Value = Text1(5).Text
            VScroll7.Value = Text1(6).Text
            VScroll8.Value = Text1(7).Text


            end_time = Timer + speed
            Do While end_time > Timer
                DoEvents
            Loop

    K = K + 1

Loop

GoTo Finish

Com_error:
    MsgBox ("Communications problem encountered")
```

```
Finish:
    Text2.Text = pos_index
    Text2.Refresh
End Sub
```

The playback routine works by transmitting the servo positions stored in the stored_positions array to the open Com port to which the serial servo controller board is attached. The routine transmits the sync byte, the servo number, and the position data to the PIC 16F84A. When the PIC 16F84A receives the data, it transmits the servo number and the position data back to the computer. The Visual Basic routine compares the data that the PIC 16F84A sends back to the information that it just sent. If the two are the same, then our program knows that the serial servo controller has properly received the data. If the data are not the same, our application sends it again and does another comparison of the data that the PIC sent back. This continues until the data matches, which should happen immediately or after a few attempts. The program keeps track of how many attempts it makes to send the data, and if it has not received a reply from the serial servo controller circuit after 50 attempts, then it times out and tells the user that there is a communications error. If this small piece of code were left out, the program would get stuck in an indefinite loop if the servo controller circuit was not powered up or there was a problem with the cable. In most cases, the data only need to be sent once, but these extra measures ensure that the arm doesn't make any wrong moves that might damage itself or something nearby.

Try out the completed program by pressing the F5 key. If everything is working properly, you can create an executable (.EXE) program. Click on File in the menu bar at the top of the Visual Basic development studio application, and select Make servo-controller.exe from the dropdown, as shown in **Figure 6.20**. You will be asked where you would like to save the application and have the opportunity to name the executable. When you click the Next button, the program will be compiled. If you plan on running this executable on machines that have Visual Basic 6.0 installed, then everything will run as expected. If you want to use the serial servo controller application on other computers, then you will need to create a distribution install package that bundles all of the dependent files together so that they can be installed together with the executable.

Package and Deployment Wizard

The Package and Deployment Wizard tool is included with the Microsoft Visual Studio package. This tool makes it easy to create installation packages for the applications you develop so that they can be distributed, along with all of the files needed for your application to function properly. Start the Package and Deployment Wizard from programs\Microsoft Visual Studio 6.0\Microsoft Visual

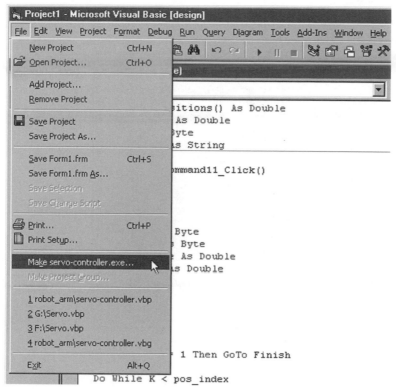

■ **FIGURE 6.20** *Making the servo controller application.*

Studio 6.0 tools\Package and Deployment Wizard. The first window you see will look similar to **Figure 6.21**. Select the Browse button to select the location where the servo controller project is stored. When you are finished selecting the project, click on the Package button.

The next window to be presented will ask you to choose a packaging script. Choose the default packaging script called Standard Setup Package 1, and then click Next.

The Included Files dialog box, like the one in **Figure 6.22**, will be shown next. The setup wizard has determined which files need to be shipped with the program when distributing the application. Notice that Visual Basic has correctly determined that our program will need the MSCOMM32.OCX file, along with others, to work properly. Click the Next button to continue.

The next window will prompt you for a location to assemble the package. If you don't have a folder ready in which to assemble the package, click the New Folder button and create one. Click Next.

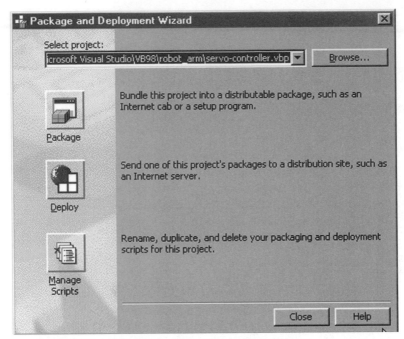

■ **FIGURE 6.21**   *The Package and Deployment Wizard.*

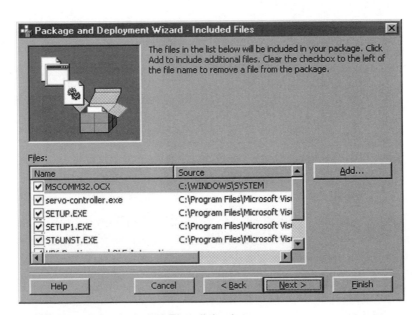

■ **FIGURE 6.22**   *Included Files dialog box.*

The Package Type dialog box, like the one shown in **Figure 6.23**, will be displayed next. Choose Standard Setup Package, and then click the Next button.

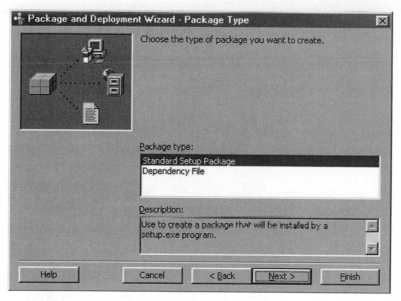

■ **FIGURE 6.23** *Package Type dialog box.*

The Cab Options dialog box will be displayed next. This will give you the choice of creating a Single Cab or Multiple Cabs. Multiple Cabs lets you determine the size so that the application can be distributed on floppy disks if desired. Choose the Single Cab option and click Next.

The Next dialog asks you to pick an installation title. Type Serial Servo Controller into the Installation Title text box, and then click Next.

Accept the default options for the Shared Files dialog box and click Next. Again, accept the default values for the Install Locations dialog and click Next.

Figure 6.24 shows the Start Menu Items dialog that will be presented. Accept the default and click Next.

The Next dialog box will ask for a script name to save the setting for this session. Accept the default Standard Setup Package 1 script name and click Next.

Visual Basic will begin creating the distribution files. Essentially, Visual Basic will create *Setup.exe*, which loads and registers all of the files within the installation into the computer. When the installation runs on the user's computer, the *Setup.exe* program will decompress each of the files and place it in the location assigned to it within the File Details dialog box. When it completes, you now have a distribution package that can be installed on other machines!

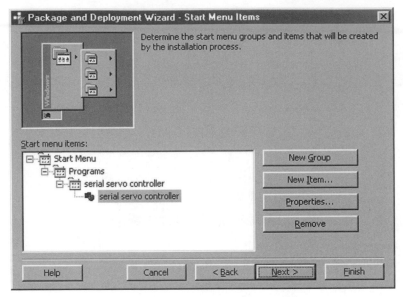

■ FIGURE 6.24 *Start Menu Items dialog box.*

Using the Serial Servo Controller Software with the Robot Arm

If all of the steps in Chapter 5 are completed, then the robot arm should be connected to the computer via the serial servo controller circuit board. **Figure 6.25** shows the connection diagram for all of the components that make up the robot arm project. Power up the controller board, and then start the serial servo controller application on your PC. Try making small movements with the first vertical slider and you will notice that the robot arm rotates on its base. Experiment with the first six sliders, but be sure to make small changes until you get a feel for how the robot arm reacts to the controls. The base, shoulder, and elbow joints actually have a lot of power and speed. Be careful not to get in the way of the robot arm when you are experimenting with it. **Figure 6.25** lists the robot arm joints and each associated vertical scroll bar of the servo controller software that is used to operate it. For the robot arm project, only the first six scroll bars are used. All eight of the scroll bars will be used for other projects in the book, but the PIC 16F84A microcontroller will be reprogrammed with the appropriate code.

To create a movement sequence, position the arm in the desired position with the scroll bars, and then click on the Store Pos button. This position will be the first in the sequence. Position the arm in the next position that is desired to carry out a task, and then click the Store Pos button. When putting a movement sequence together, it is best to move only a couple of joints per sequence step for a smoother flow of movement. When you have constructed all of the positions into a sequence, click on the Playback button and the robot arm will move through all of the

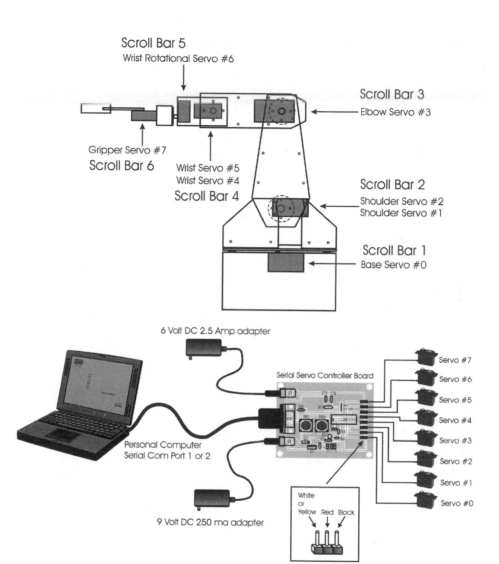

Scroll Bar 5
Wrist Rotational Servo #6

Scroll Bar 3
Elbow Servo #3

Gripper Servo #7
Scroll Bar 6

Wrist Servo #5
Wrist Servo #4
Scroll Bar 4

Scroll Bar 2
Shoulder Servo #2
Shoulder Servo #1

Scroll Bar 1
Base Servo #0

6 Volt DC 2.5 Amp adapter

Serial Servo Controller Board

Servo #7
Servo #6
Servo #5
Servo #4
Servo #3
Servo #2
Servo #1
Servo #0

Personal Computer
Serial Com Port 1 or 2

9 Volt DC 250 ma adapter

White
or
Yellow   Red   Black

■ **FIGURE 6.25**   *Robot arm connection diagram.*

recorded positions. You can then adjust the playback speed, varying the speed control scroll bar. This version of the software does not have the capability to save your sequences, but a more advanced version can be downloaded from www.thinkbotics.com.

Summary

In this chapter, the fundamentals of creating a serial communications project using Visual Basic were covered. The ability to craft your own robot control software gives you the ability to harness the computing power of your PC. Using a micro-controller in conjunction with a PC provides enormous power and flexibility to any project. The concepts in this chapter can be adapted and applied to any number of projects that require serial communications software development. It is an enormous advantage to be able to use the strengths of the microcontroller and the PC in robotic control systems and artificial intelligence applications.

7

Speech Recognition

The ability to communicate with a robot through speech is the ultimate user interface. Speech is one of the human attributes that sets us apart from the rest of the animal kingdom and is considered proof of higher intelligence. When a robot obtains the ability to recognize words, it is well on its way to becoming a true humanoid. Speech recognition provides a simple and effective means for humans to specify a task for the robot and for the robot to acquire new skills without any additional hard-coded programming. If you have ever tried to learn a second language as an adult, you know how difficult it can be to recognize the individual words and understand sentences. Teaching a machine to understand human speech can be an even harder pursuit. Today's technology has advanced to the level where low-cost speech recognition is available and can be implemented easily.

In this chapter, a speech recognition control circuit will be built that can be used for any number of projects where hands-free speech recognition is desired. The speech recognition control circuit is shown with a microphone headset in **Figure 7.1**. In our case, we will be using the speech recognition control circuit to control the robot arm by issuing spoken commands! It is quite amazing to see the robot arm responding to the spoken control words because it makes you realize that the visionary dream of truly intelligent and autonomous robots is getting closer every day.

Sensory Inc. (www.sensoryinc.com) provides an embedded speech recognition product that is perfect for a humanoid robot. The Voice Direct 364 speech recognition kit that this company produces is shown in **Figure 7.2**.

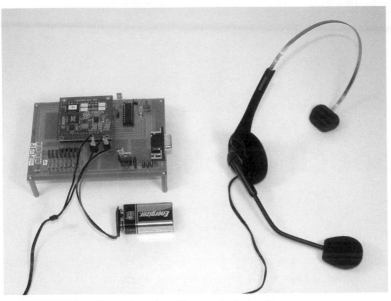

■ **FIGURE 7.1** *Speech recognition control circuit.*

122

■ **FIGURE 7.2** *Voice Direct 364 speech recognition kit.*

The Voice Direct 364 module performs speaker-dependent discrete word recognition by comparing a pattern generated in real time with previously trained word templates. The pattern generated by the Voice Direct 364 module is based on a digital reconstruction of the voice command. Each word to be recognized must first be trained. During training, the Voice Direct 364 module constructs a template representing the individual speaker's unique sound pattern for each specific word or phrase to be recognized. Templates are stored in serial EEPROM memory. During recognition, a new pattern is produced and compared to the stored templates to determine which word was spoken. The Voice Direct 364 module features integrated speech, prompting for both training and recognition operations. This allows the development of sophisticated interactive products with minimal programming. Some features of this module are:

- Speaker-Dependent and Continuous Listening speech recognition technologies.
- Minimal external components.
- Recognizes up to 60 words or phrases in slave mode, or 15 in stand-alone mode (broken into 1, 2, or 3 sets).
- Over 99% recognition accuracy with proper design.
- Phrase recognition up to 2.5 seconds.
- User-friendly speech prompting.

System Design Considerations for the Voice Direct 364 Module

If the module will be used in a system with other digital clocks (switching power supplies, LCD driver, etc.), take special care to prevent these signals from being coupled into the audio circuitry of Voice Direct 364.

With proper product construction, Voice Direct 364 meets the CE requirements for electromagnetic radiation and immunity. To minimize radiated emissions, speaker wires should be less than 3 inches long. In addition, the speaker cable and power cable should be oriented on opposite sides of the module.

Microphone Considerations

For most applications, an inexpensive omnidirectional electret capacitor microphone with a minimum sensitivity of –60 dB is adequate. In some applications, a directional microphone might be more suitable if the signal comes from a direction different from the audio noise. Since directional microphones have a frequency response that depends on their distance from the sound source, such microphones should be used with caution. For best performance, speech recognition products should be used in a quiet environment, with the speaker's mouth in close proximity to the microphone.

Important Mechanical Issues Pertaining to Microphone Assembly

In the product, the microphone element should be positioned as close to the mounting surface as possible, and it should be fully seated in any housing. There must be no airspace between the microphone element and the housing. Having such airspace can lead to acoustic resonance, which can reduce recognition accuracy.

The area in front of the microphone element must be kept clear of obstructions to avoid interference with recognition. In general, the diameter of the hole in the housing in front of the microphone should be at least 5 mm. Any necessary plastic surface in front of the microphone should be as thin as possible, being no more than 0.7 mm if possible.

The microphone should be acoustically isolated from the housing if possible. This can be accomplished by surrounding the microphone element with a spongy material such as rubber or foam. Mounting with a pliable, nonhardening adhesive is another possibility. The purpose is to prevent auditory noises, produced by handling or jarring the product, from being picked up by the microphone. Such extraneous noises can reduce recognition accuracy.

The Voice Direct 364 module performs the following operations when recognizing a word:

1. The audio signal (spoken word) is externally amplified and filtered, and then supplied to the analog inputs of the Voice Direct 364, which converts the analog waveforms to digital samples.
2. The speech signal samples are analyzed and a pattern of information representing significant speech elements is generated.
3. The Voice Direct 364 module increases or decreases the gain of the external amplifier as needed to maintain signal quality.
4. Using a neural network, the pattern is compared with previously stored template patterns; a small number of candidate templates is selected.
5. The candidate templates are further processed to determine the one template that provides the best match to the unknown pattern.
6. If the best match template gives a score above a predefined threshold, the Voice Direct 364 module chooses the word associated with that template. If no template provides a match above the threshold, a special "no match" value is chosen.

The speech recognition circuit that will be built during this chapter will be designed around the Voice Direct 364 module. Our circuit will use a PIC 16F84A microcontroller to interpret the output generated by the recognition module, and then output corresponding information by serial communications. This makes the speech recognition circuit much more flexible because it can be interfaced directly to the robot arm controller circuit via a null modem cable for hands-free control of the robot arm. Control of the robot arm by speech recognition will be covered fully in this chapter. You can also interface the speech recognition circuit directly to a personal computer (PC) or any of the popular microcontroller modules because the programmable interface controller (PIC) can be programmed to communicate using any serial protocol that you wish. The output from the module is also represented visually with light-emitting diodes (LEDs). All of the support circuitry, including function pushbuttons, a regulated 5-V direct current (DC) supply, talk LED, PIC 16F84A microcontroller, and a DB9 serial connector are included on the speech recognition motherboard that will be built. The schematic for the speech recognition motherboard circuit is shown in **Figure 7.3**. The complete parts necessary to build the speech recognition controller are listed in **Table 7.1**.

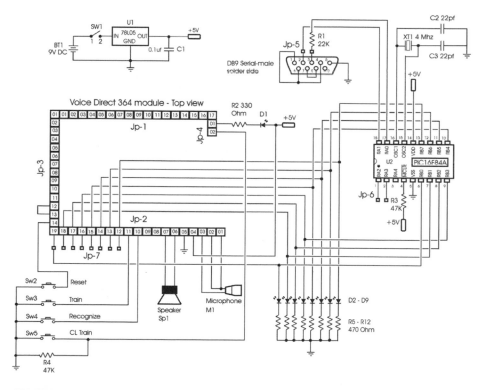

■ **FIGURE 7.3** *Speech recognition circuit schematic.*

Table 7.1 *Parts List for the Speech Recognition Circuit Board*

Part	Quantity	Description
Semiconductors		
U1	1	78L05 5V regulator
U2	1	PIC 16F84A flash microcontroller mounted in socket
D1	1	Green light-emitting diode—rectangle, 1.5 mm × 5 mm
D2–D9	8	Red light-emitting diodes—rectangle, 1.5 mm × 5 mm
Resistors		
R1	1	22 KΩ 1/4-watt resistor
R2	1	330 Ω 1/4-watt resistor
R3	1	47 KΩ 1/4-watt resistor
R4	1	47 KΩ 1/4-watt resistor
R5–R12	8	470 Ω 1/4-watt resistor
Capacitors		
C1	1	0.1 μf capacitor
C2, C3	2	22 pf
Miscellaneous		
Voice direct 364	1	Speech recognition module
JP2	1	19-post header connector—2.5 mm spacing
JP3	1	14-post header connector—2.5 mm spacing
JP4, JP5, JP6, JP8, JP9, JP10, JP11, JP12, JP13	1	2-post header connector—2.5 mm spacing
JP7	1	8-post header connector—2.5 mm spacing
SW1	1	SPST power switch
SW2–SW5	4	Momentary contact pushbuttons
SUB1	1	DB9—female 9-pin serial connector
XT1	1	4 MHz crystal
Speaker	1	2-1/4 inch—8 Ω speaker
Microphone	1	Microphone element—omnidirectional electret capacitor
Headset	1	Can be used instead of the speaker and microphone
Battery clip	1	9-V type
Standoffs with mounting screws	4	1-1/4 inch in length
DB9 male 9-pin serial connector	2	Connectors to fabricate a null modem cable to connect the speech recognition board to the serial servo controller board of the robot arm
2-connector female header	3	2.5 mm spacing
Hookup wire	4 inches	2-strand connector wire
Hookup wire	16 inches	3-strand connector wire
Heat shrink tubing	4 inches	Protect solder joints on connector wires
Printed circuit board	1	See details in chapter

Creating the Speech Recognition Controller Printed Circuit Board

To fabricate the speech recognition controller printed circuit board (PCB), photocopy the artwork in Figure 7.4 onto a transparency. Make sure that the photocopy is the exact size of the original. For convenience, you can download the file from the author's Web site, www.thinkbotics.com, and simply print the file onto a transparency using a laser or ink-jet printer with a minimum resolution of 600 dots per inch (dpi). After the artwork has been successfully transferred to a transparency, use the techniques outlined in Chapter 2 to create a board. A 4 × 6-inch presensitized positive copper board is ideal. When you place the transparency on the copper board, it should be oriented exactly the same as in Figure 7.4.

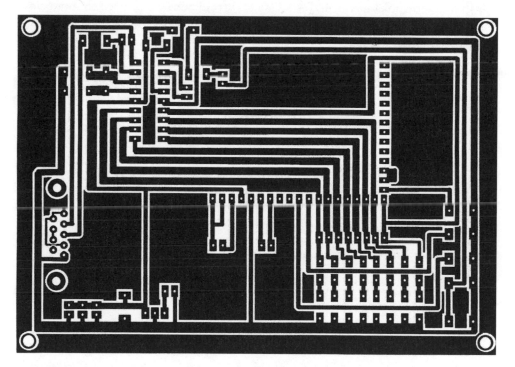

■ **FIGURE 7.4**   *PCB foil pattern for the speech recognition circuit board.*

Circuit Board Drilling and Parts Placement

Use a 1/32-inch drill bit to drill all of the component holes on the PCB. Drill the holes for the voltage regulator (U1) with a 3/64-inch drill bit. Use Table 7.1, Figure 7.3, and Figure 7.5 to place the parts on the component side of the circuit board. Note that a vertical line on each LED in Figure 7.5 indicates the position of the cathode (–) for each of the LEDs. The PIC 16F84A microcontroller (U2) is mounted in an 18-pin IC socket. The 18-pin socket is soldered to the PCB, and the PIC will be inserted after it has been programmed. Use a fine-toothed saw to cut the board along the guidelines, and drill the mounting holes on the corners using a 5/32-inch drill bit. Use four 1 1/4-inch standoffs to mount the board. Locate the Voice Direct 364 module and plug it into jumpers 2, 3, and 4 on the circuit board by lining up the left and bottom connectors with jumpers 3 and 2. Solder the 9-V battery strap to a 2-connector female header, and then connect it to Jp-10, with the positive wire to the right. Solder two 3-inch wires to the 8-ohm speaker, and then solder the wires to a 2-connector female header. Connect the speaker header to Jp-8. Solder two 1-inch wires to the electret microphone, and then solder the wires to a 2-connector female header. Connect the microphone header to Jp-9. If you prefer to use a microphone headset with a speaker earpiece, then connect the speaker to Jp-8 and the microphone to Jp-9. The headset eliminates the need to stand close to the microphone and cuts out a lot of background noise, giving a higher word recognition rate.

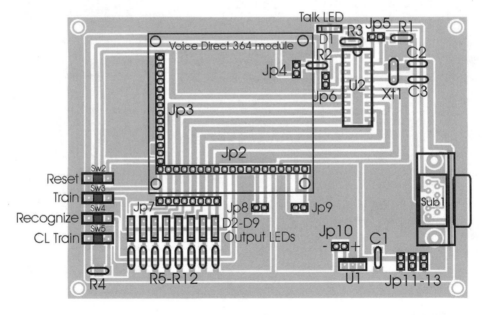

■ **FIGURE 7.5** *Parts placement diagram for the speech recognition circuit board.*

If you look at the schematic for the speech recognition board, you will notice that is has been configured for single word continuous listening mode, as outlined in the Voice Direct 364 manual. This means that the module will continuously listen for a single key word. After successful recognition of the trigger continuous listening (CL) word, there is a three-second window during which the Voice Direct 364 module is listening for a speaker-dependent (SD) word (if any are trained). If no SD words are trained, then a successful recognition of the CL word will cause the outputs to behave as if the first SD word was recognized (OUT1 would toggle high). This feature is useful if only a single command is required, such as a light switch. Note that the Voice Direct 364 module does not need to wait the full three seconds if it hears your command. Once it detects the silence at the end of your utterance, recognition processing will begin immediately.

When a trained sequence of words (CL word + SD word) is recognized, the appropriate Output Pin(s) will pulse high for one second, as listed in **Table 7.2**. The "A" indicates that the outputs are Active-High. Because the outputs only pulse high for one second, our circuit uses a microcontroller to monitor the outputs and then send out serial information about the recognized word sequence. In the stand-alone mode that we are using, 15 SD words can be trained.

Table 7.2   *Voice Direct 364 Outputs*

Recognition Word	Out1	Out 2	Out 3	Out 4	Out 5	Out 6	Out 7	Out 8
CL + SD Word 1	A							
CL + SD Word 2		A						
CL + SD Word 3			A					
CL + SD Word 4				A				
CL + SD Word 5					A			
CL + SD Word 6						A		
CL + SD Word 7							A	
CL + SD Word 8								A
CL + SD Word 9	A							A
CL + SD Word 10		A						A
CL + SD Word 11			A					A
CL + SD Word 12				A				A
CL + SD Word 13					A			A
CL + SD Word 14						A		A
CL + SD Word 15							A	A

Now that the voice recognition circuit is complete, it is time to train the Voice Direct module to recognize words. In the mode that the circuit has been configured, separate buttons are used to train SD words (TRAIN switch) and the CL word (CL TRAIN switch). The switches that will be used are located on the left of the circuit board and are labeled in **Figure 7.5**. After training the CL word, up to 15 SD commands can be trained. We will be training the module to respond to word commands that will be used to control the robot arm that was built in Chapter 5. To train the module, power it up by attaching a 9-V battery to the battery clip. Put the headset on if you decided to go with that option. Otherwise, position yourself in front of the speech recognition circuit.

To train the CL word, follow these steps:

1. Press the CL TRAIN switch on the circuit board.
2. The Voice Direct 364 module will speak, "Say word one."
3. Speak into the microphone at a normal volume level, saying "Arm."
4. The Voice Direct 364 module will speak, "Repeat."
5. Speaking into the microphone say "Arm."
6. The Voice Direct 364 module will then say "Accepted."

To start training the 15 SD words, follow these steps:

1. Press the TRAIN switch.
2. The Voice Direct 364 module will speak, "Say word one."
3. Speak into the microphone at a normal volume level saying, "Base left."
4. The Voice Direct 364 module will speak, "Repeat."
5. Speak into the microphone at a normal volume level saying, "Base left."
6. The Voice Direct 364 module will then say, "Accepted."

Continue training the rest of the words in the list below by pressing the TRAIN switch and following the same procedure listed above. The next time the TRAIN switch is pressed the module will speak, "Say word two," until it gets to word 15.

Continuous listening word 1: Arm

Word 1: Base left

Word 2: Base right

Word 3: Shoulder up

Word 4: Shoulder down

Word 5: Elbow up

Word 6: Elbow down

Word 7: Wrist up

Word 8: Wrist down

Word 9: Left wrist

Word 10: Right wrist

Word 11: Gripper open

Word 12: Gripper close

Word 13: Slow

Word 14: Fast

Word 15: Memory one

If a problem is encountered when training words, the Voice Direct module will say, "Training Error." This usually occurs when repeating the word that is being trained is not similar enough to the first time that you said it. If you don't talk loud enough during word training, the module will say, "Please talk louder." Trying to train two similar SD words will result in a "too similar" error, and you will be prompted to say the word again. During SD training, each SD word is checked for similarity to the other words in the set.

Key Considerations for Successful Voice Recording

The equipment used to train the voice recordings should match the equipment used during recognition. Differences in microphone, microphone housing, etc., will adversely affect recognition. The conditions and environment in which the voice recordings are made should reflect the conditions and environment in which the end product will be used. Some of the factors that can affect recognition reliability are listed below.

- **Distance.** The distance of the microphone from the speaker's mouth must be the same during recording and during end-product use. For example, a doll is typically held within arm's length, so the voice recording microphone should be held accordingly.

- **Natural Voices.** Subjects should speak in their normal voice and should be discouraged from sounding different by imitating a foreign accent or using any unnatural intonation. They should be prompted by means of some nonverbal source (pictures or flashcards, for example), so as not to unconsciously mimic the voice of the person doing the prompting.

- **Physical States.** Physical states should be considered. For example, in collecting voice recordings for an exercise machine, it is strongly recommended to record people who are out of breath.

- **Emotional States.** Emotional states should be considered. Will the end users be relatively quiet and calm (say, for an office product), or loud and excited?

- **Environment/Background Noise.** Environmental noise must be considered. Voice recordings should ideally be made in an environment similar to the one in which the end product will be used. The speech signal must be prominent relative to background noise, and there should not be any abrupt, loud noises. Voice recordings should not be made in a soundproof room. These rooms lend an unnatural background silence to the recordings, which does not reflect the real-world environment in which the end product will be used.

Testing the Speech Recognition Circuit

After the CL word and the 15 SD words have been trained, push the RECOGNIZE button. Say the word "Arm." If the word was recognized, the Talk LED (see **Figure 7.5**) will quickly blink off and on. You now have three seconds to say one of the SD words. Say the word "Base Left." If the word is recognized, the Voice Direct module will say "One," and the red LED on the far right will turn on for one second. Get familiar with the module by saying the word "Arm," followed by each of the previously trained words. If you want to erase the template and re-record the words or use your own words to control the arm, then press the TRAIN and RECOGNIZE buttons simultaneously for a second. Voice Direct will erase all of the trained templates and will then say "Memory erased."

Programming the PIC 16F84A Microcontroller

Because the Voice Direct module can recognize 15 words in stand-alone mode, but has only eight outputs, the PIC 16F84 will be used to decode the output combinations listed in **Table 7.2**. Once the outputs have been decoded, the PIC will generate a serial output signal on Port A, pin 1. The serial output will include a sync byte (255), and then the number that corresponds to the recognized word (1 through 15) at 2400 baud. The PIC 16F84A microcontroller on the serial servo controller circuit board, used to control the robot arm, will also be reprogrammed to accept this information so that it can move the arm according to the commands that it receives. Compile the voice-dir.bas program listed in **Program 7.1**. Program the PIC 16F84A with the voice-dir.hex file listed in **Program 7.2**. When the PIC 16F84A has been programmed, insert it into the 18-pin socket on the speech recognition circuit board with pin 1 to the top left of the socket, as illustrated in **Figure 7.5**.

■ **PROGRAM 7.1**   *voice-dir.bas program listing.*

```
'_____

' Name      : voice-dir.bas
' Compiler  : PicBasic Pro - MicroEngineering Labs
' Notes     : Voice Direct 364 to serial interface
'_____


TRISB = %11111111
TRISA = %00000001

DEFINE OSC 4

include "modedefs.bas"
```

```
Baud        CON N2400
Com_In      VAR PORTA.0
Com_Out     VAR PORTA.1
delay       VAR BYTE
delay = 0

Serout com_out,Baud,[255,11] ' wake up serial servo controller

Start:

    If PortB = 128 then Serout com_out,Baud,[255,1]
    If PortB = 64 then Serout com_out,Baud,[255,2]
    If PortB = 32 then Serout com_out,Baud,[255,3]
    If PortB = 16 then Serout com_out,Baud,[255,4]
    If PortB = 8 then Serout com_out,Baud,[255,5]
    If PortB = 4 then Serout com_out,Baud,[255,6]
    If PortB = 2 then Serout com_out,Baud,[255,7]
    If PortB = 1 then Serout com_out,Baud,[255,8]
    If PortB = 129 then Serout com_out,Baud,[255,9]
    If PortB = 65 then Serout com_out,Baud,[255,10]
    If PortB = 33 then Serout com_out,Baud,[255,11]
    If PortB = 17 then Serout com_out,Baud,[255,12]
    If PortB = 9 then delay = 20
    If PortB = 5 then delay = 0
    If PortB = 3 then Serout com_out,Baud,[255,15]

    pause delay

goto start
```

■ **PROGRAM 7.2**   *voice-dir.hex file listing.*

```
:10000000632892002208840009309300031000D2019
:10001000920C930B072803140D2884139F1D1C2892
:100020000000820041F1D2006800084170008200 4FB
:100030000031C2006800027280082004031C20063B
:100040001F19200680008417200980052728 1F0D0E
:100050006398C0030208D008C0A302000004A28A0
:100060000000308A000C0882070134753403341534DB
:100070000000343C340C34D9348F018E00FF308E07AD
:100080000031C8F07031C5E2803308D00DF304A20DD
:100090003E288D01E83E8C008D09FC30031C53285E
:1000A0008C07031850288C0764008D0F50280C18FB
```

```
:1000B00059288C1C5D2800005D280800831303135 9
:1000C0008312640008008316FF30860001308500 2B
:1000D0008312A4010530A2000230A00004309F006A
:1000E000FF3001200B30012064000608803C031D16
:1000F00083280530A2000230A00004309F00FF30AA
:1001000001200130012064000608403C031D9228B4
:100110000530A2000230A00004309F00FF30012013
:100120000023001206400060 8203C031DA128053090
:10013000A2000230A00004309F00FF3001200330F5
:1001400001206400060 8103C031DB0280530A20001
:100150000230A00004309F00FF3001200430012055
:1001600064000608083C031DBF280530A2000230C9
:10017000A00004309F00FF300120053001206 40002
:100180000608043C031DCE280530A2000230A00062
:1001900004309F00FF30012006300120640006087 3
:1001A000023C031DDD280530A2000230A00004300F
:1001B0009F00FF300120073001206400060 8013C49
:1001C000031DEC280530A2000230A00004309F007F
:1001D000FF300120083001206400060 8813C031D27
:1001E000FB280530A2000230A00004309F00FF3041
:1001F0000001200930012064000608413C031D0A2942
:100200000530A2000230A00004309F00FF30012022
:100210000A3001206400060 8213C031D192905301D
:10022000A2000230A00004309F00FF3001200B30FC
:1002300001206400060 8113C031D28290530A20096
:100240000230A00004309F00FF3001200C3001205C
:1002500064000608093C031D2F291430A400640023
:100260000608053C031D3529A40164000608033C6B
:10027000031D44290530A2000230A00004309F0075
:0E028000FF3001200F30012024083C2074289C
:02400E00F53F7C
:00000001FF
```

The next step is to reprogram the PIC 16F84A microcontroller situated on the serial servo controller circuit board to which the robot arm is connected. Compile the arm-voice.bas program listed in **Program 7.3**, and then program the PIC 16F84A with the arm-voice.hex file listed in **Program 7.4**. When the PIC 16F84A has been programmed, insert it into the 18-pin socket on the serial servo controller circuit board with the notch located toward the two push-button switches (see **Figure 5.59** in Chapter 5).

```
'_____

' Name     : Arm-voice.bas
' Compiler : PicBasic Pro - MicroEngineering Labs
' Notes    : Robot arm / Voice direct 364
'_____


TRISB = %00000000
TRISA = %11111111


DEFINE OSC 4


include "modedefs.bas"


Led          VAR PORTA.4
Baud         CON N2400
Com_In       VAR PORTA.0
Com_Out      VAR PORTA.1
Control      VAR BYTE
increment    VAR BYTE
S0           VAR BYTE
S1           VAR BYTE
S2           VAR BYTE
S3           VAR BYTE
S4           VAR BYTE
S5           VAR BYTE
S6           VAR BYTE
S7           VAR BYTE


LOW PORTB.0
LOW PORTB.1
LOW PORTB.2
LOW PORTB.3
LOW PORTB.4
LOW PORTB.5
LOW PORTB.6
LOW PORTB.7
HIGH Led


S0 = 127
S1 = 111
S2 = 144
```

```
S3 = 65
S4 = 127
S5 = 127
S6 = 127
S7 = 127
increment = 1
control = 11

Start:

    SERIN Com_In,Baud,7,Set_Pos,[255],Control

    '————— Base —————
       if Control = 1 then
          S0 = S0 - increment
          If S0 < 40 then S0 = 40
       endif
       if Control = 2 then
          S0 = S0 + increment
          If S0 > 230 then S0 = 230
       endif
    '————— Shoulder —————
       if Control = 3 then
          S1 = S1 - increment
          If S1 < 40 then S1 = 40
          S2 = 254 - S1
       endif
       if Control = 4 then
          S1 = S1 + increment
          If S1 > 230 then S1 = 230
          S2 = 254 - S1
       endif
    '————— Elbow —————
       if Control = 5 then
          S3 = S3 + increment
          If S3 > 230 then S3 = 230
       endif
       if Control = 6 then
          S3 = S3 - increment
          If S3 < 40 then S3 = 40
       endif
    '————— Wrist —————
       if Control = 7 then
```

```
            S4 = S4 - increment
            If S4 < 40 then S4 = 40
            S5 = 254 - S4
        endif
        if Control = 8 then
            S4 = S4 + increment
            If S4 > 230 then S4 = 230
            S5 = 254 - S4
        endif
'————— Wrist —————
        if Control = 9 then
            S6 = S6 + increment
            If S6 > 230 then S6 = 230
        endif
        if Control = 10 then
            S6 = S6 - increment
            If S6 < 40 then S6 = 40
        endif
'————— Gripper —————
        if Control = 11 then
            S7 = S7 - increment
            If S7 < 10 then S7 = 10
        endif
        if Control = 12 then
            S7 = S7 + increment
            If S7 > 230 then S7 = 230
        endif


Set_Pos:

        PULSOUT PORTB.0,S0
        PULSOUT PORTB.1,S1
        PULSOUT PORTB.2,S2
        PULSOUT PORTB.3,S3
        PULSOUT PORTB.4,S4
        PULSOUT PORTB.5,S5
        PULSOUT PORTB.6,S6
        PULSOUT PORTB.7,S7


GOTO Start
```

■ PROGRAM 7.4   *arm-voice.hex file listing.*

```
:100000008E28A00062200C080D0403198928832083
:100010008413200880066400D280E288C0A03191A
:100020008D0F0B288006892823088C0021088D005D
:1000300001308E008F0164003C20031C2F288E0BA2
:100040001B28FF308F0703181B288C07031C8D0704
:100050031C892832308E0000308F001B28472077
:1000600008308F0048203C208E0C36288F0B322819
:10007000482003140E0808002208840020088441772
:1000800080004841300051F192006FF3E08001F1777
:100090001F0D06398C0056208D008C0A56201F1F1C
:1000A00067281F138C0002307E20672800308A00EA
:1000B0000C08820701347534033415340343C34A1
:1000C0000C34D934FF3A8417800589288D01E83E25
:1000D0008C008D09FC30031C70288C0703186D28D8
:1000E0008C0764008D0F6D280C1876288C1C7A28DC
:1000F00000007A28080003108D0C8C0CFF3E0318BA
:100100007B280C0889288C098D098C0A03198D0A13
:10011000080083130313831264000800831686010A
:10012000FF308500831206108316061083128610 96
:100130008316861083120611831606118312861108
:100140008316861183120612831606128312861 2F4
:100150008316861283120613831606138312861 3E0
:100160008316861383120516831605128312 7F30B9
:10017000A6006F30A7009030A8004130A9007F3062
:10018000AA007F30AB007F30AC007F30AD00013083
:10019000A5000B30A4000530A2000130A0000430FF
:1001A0009F000730A300A1011420031C9229FF3CEB
:1001B000031DD4281420031C9229A40064002408E1
:1001C00013C031DEC282508A60264002830260205
:1001D0000318EC282830A60064002408023C031D04
:1001E000FA282508A6076400E7302602031CFA282F
:1001F000E630A60064002408033C031D0B292508F3
:10020000A7026400283027020318082928 30A70015
:100210002708FE3CA80064002408043C031D1C2998
:100220002508A7076400E7302702031C1929E630D8
:10023000A7002708FE3CA80064002408053C031D15
:100240002A292508A9076400E7302902031C2A2966
:10025000E630A90064002408063C031D382925085F
:10026000A9026400283029020318382928 30A9007F
:1002700064002408073C031D49292508AA026400DC
:1002800028302A02031846292830AA002A08FE3CF2
:10029000AB0064002408083C031D5A292508AA075E
```

:1002A0006400E7302A02031C5729E630AA002A0816
:1002B000FE3CAB0064002408093C031D68292508A6
:1002C000AC076400E7302C02031C6829E630AC0060
:1002D000640024080A3C031D76292508AC0264004A
:1002E00028302C02031876292830AC00640024083A
:1002F0000B3C031D84292508AD0264000A302D0241
:10030000031884290A30AD00640024080C3C031D46
:1003100092292508AD076400E7302D02031C9229BD
:10032000E630AD0026088C008D01063084000130D7
:1003300012027088C008D0106308400023001204
:1003400028088C008D010630840004300120290823
:100350008C008D0106308400083001202A088C00B2
:100360008D0106308400103001202B088C008D0197
:100370000630840020300120 2C088C008D010630CE
:100380008400403001202D088C008D01063084004F
:060390008030012 0CB28A3
:02400E00F53F7C
:00000001FF

To connect the speech recognition circuit to the serial servo controller circuit, a crossed (null modem) serial cable will need to be fabricated. Locate two DB9 male, 9-pin serial connectors and the 16-inch, 3-strand connector wire. Solder the cable together according to the wiring diagram in **Figure 7.6**. Connect the speech recognition circuit to the serial servo controller using the cable that was just created. **Figure 7.7** shows the two circuits, along with the robot arm, integrated together.

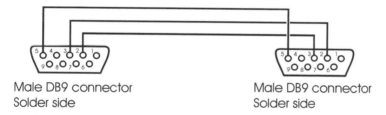

Male DB9 connector
Solder side

Male DB9 connector
Solder side

■ **FIGURE 7.6**   *Crossed (null modem) serial cable wiring diagram.*

Controlling the Robot Arm with Spoken Commands

The completed integration of the robot arm, serial servo controller, and the speech recognition circuit is shown in **Figure 7.8**.

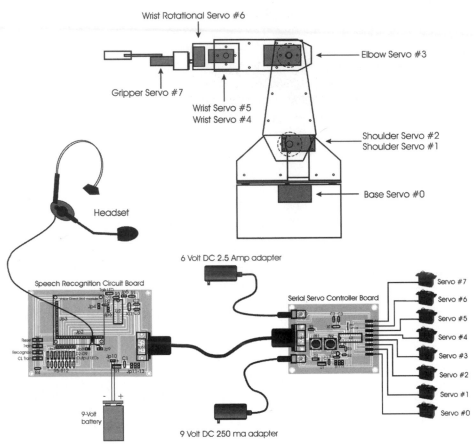

Wrist Rotational Servo #6

Elbow Servo #3

Gripper Servo #7

Wrist Servo #5
Wrist Servo #4

Shoulder Servo #2
Shoulder Servo #1

Headset

Base Servo #0

6 Volt DC 2.5 Amp adapter

Servo #7
Servo #6
Servo #5
Servo #4
Servo #3
Servo #2
Servo #1
Servo #0

Speech Recognition Circuit Board

Serial Servo Controller Board

9-Volt
battery

9 Volt DC 250 ma adapter

■ **FIGURE 7.7** *Robot arm, serial servo controller, and speech recognition circuits connection diagram.*

Apply power to the speech recognition circuit first, and then to the serial servo controller board. The order in which the circuits are powered up is important because if the serial servo controller circuit is powered up before the speech recognition circuit, there is a chance that the speech recognition template could be erased and would have to be trained again. When the speech recognition circuit is powered up, most of the red LEDs will light up briefly, and then the green talk LED will turn on and a beep will be generated. The Voice Direct module is ready to recognize the CL word when the green talk LED is on. In our case, the CL word is "Arm." Say the word "Arm" into the microphone. When the CL word is recognized, the green talk LED will blink off and on quickly. At this point, you have three seconds to say one of the 15 SD words. Say the words "Base left" within the three-second time frame after the CL word is recognized. You will notice that if the word is recognized immediately, the module will not make you wait the remaining

time of the three-second recognition window. As soon as the word is recognized, the Voice Direct module will say "Word one," and then light up the first LED for one second. During this time, the green talk LED will be off, and the module will not be listening for the CL word. For our application, this doesn't matter because the robot arm will be carrying out the last command that was issued during that time. The robot arm will rotate to the left at a reasonable speed for one-second. Try controlling the arm by issuing each of the commands that were trained earlier in the chapter. For more precise control of the arm, say the SD word "Slow," and then continue speaking commands. To go back faster, but have less precise control, say the SD word "Fast." The only complaint I have with the Voice Direct module is that the CL word has to be said before one of the 15 SD command words can be recognized. Remember that you have to say the word "Arm" before saying each of the SD command words to control the arm.

The last SD word that was trained was "Memory one." You can program whatever you want for this command, such as a macro made up of a sequence of movements. If the arm will be used by a handicapped person, then a number of macros could be defined. A useful sequence that might be used often could be a set of movements for drinking a cup of coffee.

Summary

The speech recognition control circuit that was developed in this chapter is versatile enough to be used for any number of projects where hands-free operation is needed. You can easily reprogram the PIC 16F84A to transmit whatever information you would like to use for your projects. The ability to talk to your robots and see them respond is very exciting! We are now getting closer to the dream of independent humanoids that can understand and respond to our spoken commands.

8

Build Your Own Humanoid Head

The humanoid head and face that will be built in this chapter will be used as a method of conveying emotions through facial expression. This kind of machine-to-human interface serves the purpose of making the experience of interacting with machines much more natural and enjoyable. My first encounter with a robotic-like head was experienced while visiting the Harry Houdini museum in Niagara Falls, Ontario as a child during the 1970s. Positioned at the front of the museum was a darkened glass case containing a white mannequin head wearing a hairpiece. When a person approached the exhibit, a film projector would display a talking face on the white mannequin head, giving it the three-dimensional effect of an actual human. It was as if Houdini was brought back from the dead and talking to his audience. As a child, this was very fascinating, and at the same time unnerving. I imagined that much more advanced machines would be coming soon.

The fact that the Houdini head disturbed me as a child may have something to do with what Masahiro Mori calls the "Uncanny Valley." This highly influential paper, published in the 1970s, shows insight into the relationship between robotic design and human psychology. Mori's idea states that if you plot the similarity of machines to humans on the x-axis of a graph against emotional reaction on the y-axis, you'll find that something strange happens as the robots move towards becoming perfectly lifelike. The curve rises steadily, with emotional acceptance growing, as robots become more human-like. But at a certain point, just before looking completely real, the curve plunges down, through neutrality and into revulsion, before rising again to a second peak of acceptance that corresponds with 100 percent human-like. This chasm, known as Mori's "Uncanny Valley," represents the notion that something that is very much like a human, but slightly off, will make people withdraw. The head that we are going to build, shown in **Figure 8.1**, won't even get close to plunging people into the Uncanny Valley. It will defi-

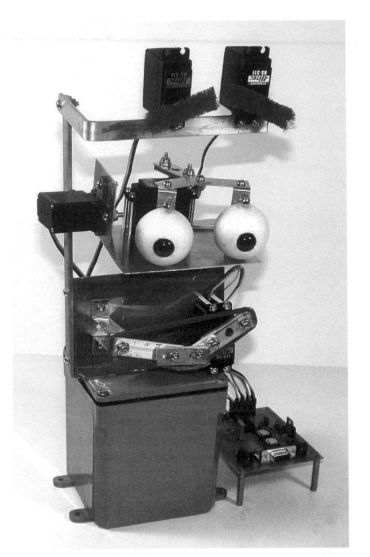

■ **FIGURE 8.1** *Humanoid head used to communicate nonverbal emotions.*

nitely be able to communicate the basic human emotions that can be conveyed using the face, but in a more simplistic way. We will also describe how a speech synthesis circuit can be added to the head so that it can have the ability to talk.

Mechanical Construction

The humanoid head project will begin with the mechanical construction. The parts needed to build the mechanical portion of the head are listed in **Table 8.1.**

TABLE 8.1 *Parts List for the Humanoid Head Mechanical Construction*

Parts	Quantity
6/32 × 1/2-inch machine screws	26
6/32 × 1 1/2-inch machine screws	10
6/32 locking nuts	36
1/16-inch thick flat aluminum stock	22 × 9-inch piece
1/2 × 1/8-inch aluminum stock	30 inches
3/4 × 3/4-inch angle aluminum	2 inches
1/4-inch diameter hollow aluminum tubing	4 inches
Plastic utility box: 4 inches wide × 4 inches in length × 4 1/2-inches deep—The Home Depot	1
1 1/2-inch diameter Styrofoam ball—craft store	2
Glass iris—craft store	2
Hair elastic—red	1
Hitec standard servo and hardware HS-311	5
Servo connector extension wires: 12 inches in length	2
Thick red pipe cleaner	1
Serial servo controller circuit board (built in Chapter 5)	1
Serial cable	1

The first part of the head that will be constructed will be the mouth. Start by cutting a piece of 1/16-inch flat aluminum stock to a size of 5 × 3 inches. Drill and bend the piece according to the dimensions shown in **Figure 8.2**. This piece is the

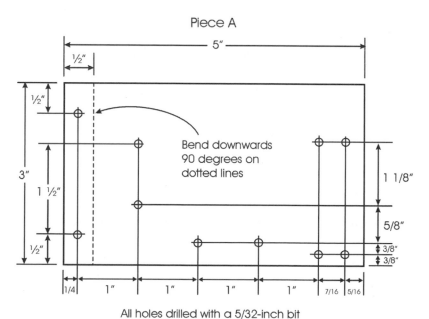

All holes drilled with a 5/32-inch bit

■ **FIGURE 8.2** *Cutting, drilling, and bending diagram for the mouth platform.*

platform to which the mouth mechanism and servo will be mounted, and is identified as piece A. Using the 1/16-inch thick flat aluminum, cut, drill, and bend piece B, as shown in **Figure 8.3**. Use the 1/2-inch wide × 1/8-inch thick aluminum stock to cut and drill pieces C, D, and E to the dimensions shown in **Figure 8.3**. Use a small metal file to elongate the holes in Piece D.

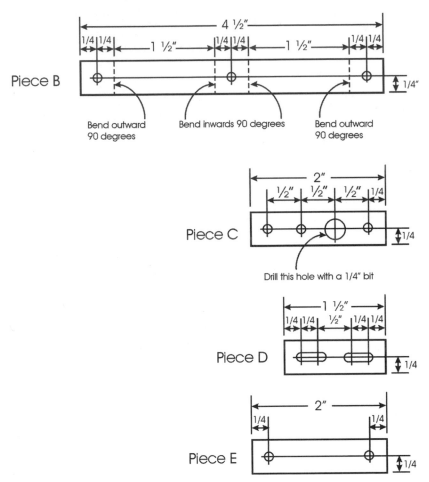

■ **FIGURE 8.3** *Cutting, drilling, and bending diagram for humanoid head parts.*

Refer to **Figure 8.4** when assembling the mouth section of the head. Secure one of the Hitec HS-311 standard servos to piece A, using four 6/32 × 1 1/2-inch machine screws and locking nuts. Prepare a servo horn according to the instruction in Chapter 5 (**Figure 5.47** through **Figure 5.50**) so that piece C can be attached. Mount the servo horn assembly to the servo with the mounting screw so that it is situated in its middle position. Mount piece B to piece A with two 6/32 × 1/2-inch machine screws and locking nuts. Mount piece E to piece B with a 6/32

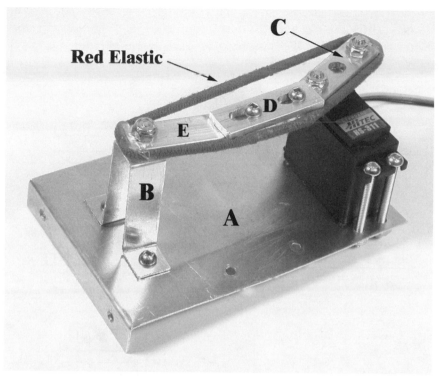

× 1/2-inch machine screw and locking nut, with two 6/32 nylon washers separating the pieces. Tighten the machine screw with enough pressure to hold the part securely in place, but to allow it to move freely. Mount piece D to piece E and piece C with two 6/32 × 1/2-inch machine screws and locking nuts. Again, tighten the machine screws with enough pressure to hold the parts securely in place, but to allow them to move freely. Move the servo horn by hand to make sure that none of the parts bind. You may have to file the corner edges off of piece D if you find that they catch on the locking nut situated on the servo horn. Place the red hair elastic around pieces C, D, and E. Use hot glue to fix the elastic to the bottom and top of piece D.

Cut a piece of 3/4 × 3/4-inch angle aluminum to a length of 2 inches, and then drill according to the dimensions shown for piece F in **Figure 8.5**. Locate the lid for the utility box and position piece F at the middle. Mark the lid in the location of the mounting holes in piece F. Drill the holes in the lid with a 5/32-inch bit, and then mount piece F to the lid with two 6/32 × 1/2-inch machine screws and locking nuts. Mount the lid on the utility box, as shown in **Figure 8.6**. Attach piece A to piece F using two 6/32 × 1/2-inch machine screws and locking nuts, as pictured in **Figure 8.7**.

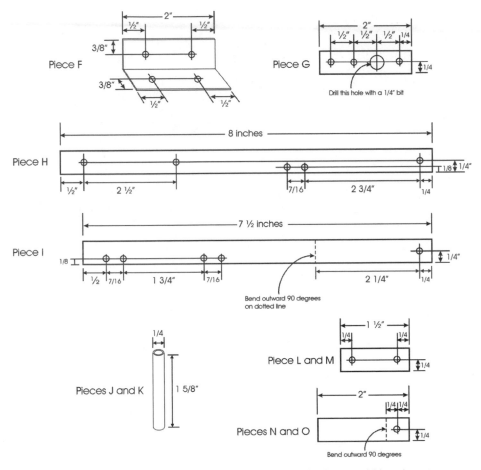

■ **FIGURE 8.5**   *Cutting, drilling, and bending diagram for humanoid head parts.*

Using 1/2 × 1/8-inch aluminum stock, cut, drill, and bend pieces H and I according to the dimensions shown in **Figure 8.5**. Fabricate pieces G, L, M, N, and O using 1/16-inch thick aluminum according to the dimensions shown in **Figure 8.5**. Locate the 1/4-inch aluminum tubing and cut two pieces, labeled J and K, to a length of 1-5/8 inches, as illustrated in **Figure 8.5**. Cut a piece of 1/16-inch flat aluminum to a size of 6 × 3-inches. Use **Figure 8.8** as a cutting, drilling, and bending guide to fabricate the eye mechanism mounting piece labeled P.

■ **FIGURE 8.6** *Piece F mounted to the utility box lid.*

■ **FIGURE 8.7** *Piece A attached to piece F.*

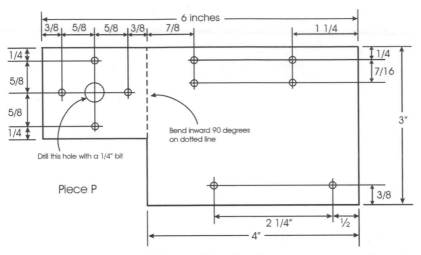

■ **FIGURE 8.8** *Cutting, drilling, and bending diagram for eye mounting piece P.*

Attach piece H to piece I using one 6/32 × 1/2-inch machine screw and a locking nut, as pictured in **Figure 8.9**. Attach a standard servo to piece H, with the output shaft positioned closest to piece H, using two 6/32 × 1/2-inch machine screws and locking nuts. This servo will be used to move the eyes up and down vertically. Attach two standard servos to piece I, with the output shafts located closest to piece I, using two 6/32 × 1/2-inch machine screws and locking nuts, as pictured in **Figure 8.9**. Attach a straight servo horn to each of the two servos mounted to piece I. Make sure that each of the servo shafts are positioned in their middle positions when the horns are horizontal with piece I, and then secure in place with the mounting screw. Cut two pieces of the red pipe cleaner to a length of 2 inches each. Use hot glue to fix one of the pipe cleaner pieces to the servo horn of the left servo attached to piece I. Fix the second pipe cleaner piece to the servo horn of the right servo attached to piece I. These will act as eyebrows.

Locate the two Styrofoam balls and drill a 1/4-inch hole through each one at the center. Insert aluminum tube piece J into the hole in one of the Styrofoam balls so that it goes all the way through and protrudes the same amount on each side. Use hot glue to secure it in place. Do the same for the second Styrofoam ball and piece K. Locate one of the glass irises and place it in the center of one of the Styrofoam balls on the opposite side as the aluminum tube piece. Use a pencil to trace around the iris, and then use a utility knife to cut enough Styrofoam out of the area to allow the iris to sit in the hole. Do the same for the second Styrofoam ball. Use hot glue to secure each glass iris into the holes that were just cut. Refer to **Figure 8.10** to see how the eyes are assembled. Locate piece N and piece O. Use hot glue to attach each one to a Styrofoam ball on the side opposite to the iris. The longest section of each piece is attached to the ball, and the short section with the hole should be even with the top of the ball, as shown in **Figure 8.10**. Use four 6/32 ×

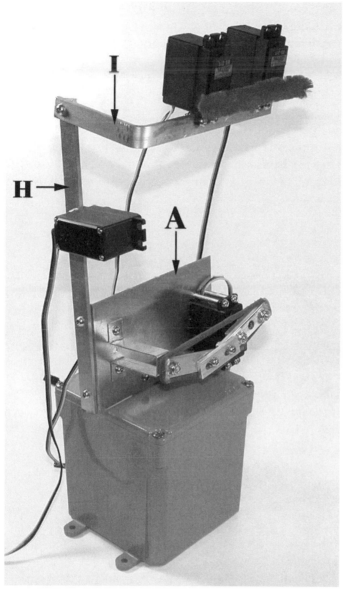

■ FIGURE 8.9 *Pieces H and I attached to piece A.*

1 1/2-inch machine screws and locking nuts to attach the servo to piece P with the output shaft located at the center.

Prepare a servo horn according to the instructions in Chapter 5 (**Figure 5.47** through **Figure 5.50**), and then attach piece G using two 6/32 × 1/2-inch machine screws and locking nuts, as pictured in **Figure 8.10**. Position the servo output shaft to its middle position, and then secure the servo horn in place so that it is facing toward the front of piece P, as shown in Figure 8.10. Secure each eyeball in place on piece P using two 6/32 × 1 1/2-inch machine screws and locking nuts. Tighten the locking nuts with enough force to hold the eyes in place, but to allow them to rotate freely. Attach mechanical linkage pieces L and M to pieces G, N, and O using three 6/32 × 1/2-inch machine screws and locking nuts, as shown in **Figure 8.10**. Secure each of the locking nuts with enough force to allow the parts to move freely. Attach the eye mechanism vertical movement servo horn to piece P using four 6/32 × 1/2-inch machine screws and locking nuts, as shown in **Figure 8.10**. The mounting holes in the servo horn will need to be drilled using a 5/32-inch drill bit. Attach the eye mechanism to the servo mounted on piece H so that it has full motion upwards and downwards from the center position, as shown in **Figure 8.11**.

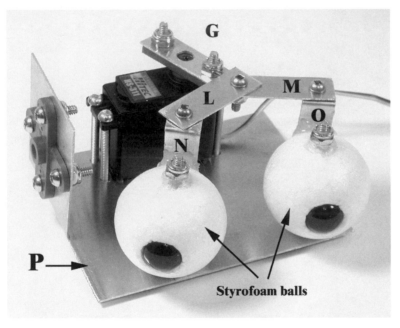

■ **FIGURE 8.10** *Eye mechanism assembly.*

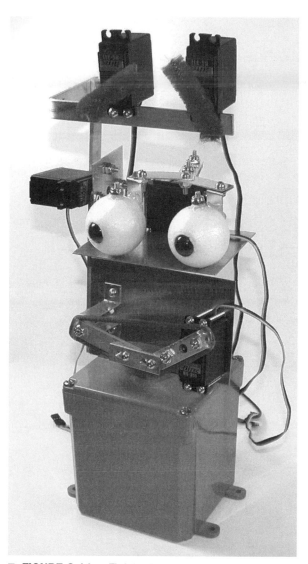

■ **FIGURE 8.11**   *Finished mechanical construction of the humanoid head.*

Interfacing the Humanoid Head

Now that the mechanical construction is complete, the humanoid head will be attached to the servo controller circuit that was built in Chapter 5 to control the robot arm. Locate the serial servo controller circuit that was built in Chapter 5 and connect each of the head servos to the servo connectors indicated in **Figure 8.12**. Connect the circuit to a personal computer (PC) via a serial cable, also as shown in **Figure 8.12**. The humanoid head connected to a PC is pictured in **Figure 8.13**.

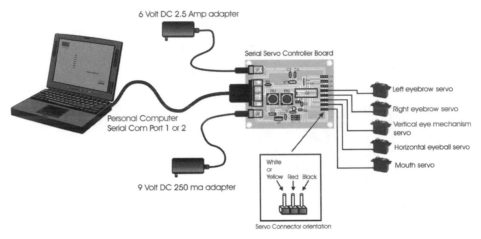

■ **FIGURE 8.12** *Humanoid head to serial servo controller connection diagram.*

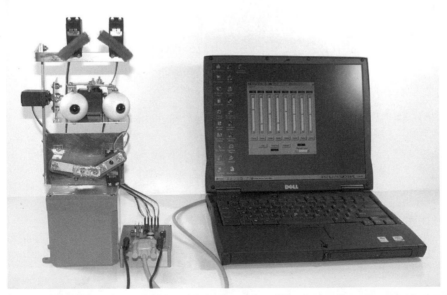

■ **FIGURE 8.13** *Humanoid head interfaced to a PC running the servo control software.*

Before applying power to the serial servo controller circuit, the PIC 16F84 micro-controller will need to be reprogrammed with the humanoid head software. Compile the PicBasic Pro program called h-head.bas, listed in **Program 8.1**, and then program the PIC 16F84A microcontroller with the h-head.hex file, listed in **Program 8.2**. Insert the PIC 16F84 back into the 18-pin socket on the serial servo controller circuit board, and then apply both power supplies.

■ **PROGRAM 8.1**   *h-head.bas program listing.*

```
'_____
' Name    : H-Head.bas
' Compiler: PicBasic Pro MicroEngineering Labs
' Notes   : Humanoid Head software
'_____


TRISB = %00000000
TRISA − %00001101

DEFINE OSC 4

include "modedefs.bas"

Led        VAR PORTA.4
Baud       CON N2400
Com_In     VAR PORTA.0
Com_Out    VAR PORTA.1
Control    VAR BYTE
Slider     VAR BYTE
S0         VAR BYTE
S1         VAR BYTE
S2         VAR BYTE
S3         VAR BYTE
S4         VAR BYTE
S5         VAR BYTE
S6         VAR BYTE
S7         VAR BYTE

LOW PORTB.0
LOW PORTB.1
LOW PORTB.2
LOW PORTB.3
LOW PORTB.4
LOW PORTB.5
```

```
LOW PORTB.6
LOW PORTB.7
HIGH Led

S0 = 127
S1 = 127
S2 = 127
S3 = 127
S4 = 127
S5 = 127
S6 = 127
S7 = 127

Start:

    SERIN Com_In,Baud,7,Set_Pos,[255],Slider,Control
    SEROUT  Com_Out,Baud,[Slider,Control]

    IF Slider = 0 THEN S0 = Control
    IF Slider = 1 THEN S1 = Control
    IF Slider = 2 THEN S2 = Control
    IF Slider = 3 THEN S3 = Control
    IF Slider = 4 THEN S4 = Control
    IF Slider = 5 THEN S5 = Control
    IF Slider = 6 THEN S6 = Control
    IF Slider = 7 THEN S7 = Control

Set_Pos:

    PULSOUT PORTB.0,S0
    PULSOUT PORTB.1,S1
    PULSOUT PORTB.2,S2
    PULSOUT PORTB.3,S3
    PULSOUT PORTB.4,S4
    PULSOUT PORTB.5,S5
    PULSOUT PORTB.6,S6
    PULSOUT PORTB.7,S7

GOTO Start
```

■ PROGRAM 8.2 *h-head.hex file listing.*

```
:10000000B428A00088200C080D040319AF28A920EB
:100010008413200880066400D280E288C0A03191A
:100020008D0F0B288006AF2823088C0021088D0037
:1000300001308E008F0164003C20031C2F288E0BA2
:100040001B28FF308F0703181B288C07031C8D0704
:100050000031CAF2832308E0000308F001B286D202B
:100060008308F006E203C208E0C36288F0B3228F3
:100070006E2003140E0808002208840020088417 4C
:100080008004841300051F192006FF3E080092001B
:10009000220884000930930003105320920C930B24
:1000A0004D280314532884139F1D62280008200440
:1000B0001F1D20068000841700082004031C200652
:1000C00080006E2800082004031C20061F1920064B
:1000D0008000841720098005 6E281F171F0D063920
:1000E0008C007C208D008C0A7C201F1F8D281F1304
:1000F0008C000230A4208D2800308A000C08820772
:10010000013475340334153400343C340C34D934A0
:10011000FF3A84178005AF288D01E83E8C008D09D9
:10012000FC30031C96288C07031893288C07640066
:100130008D0F93280C189C288C1CA0280000A02848
:10014000080003108D0C8C0CFF3E0318A1280C082E
:10015000AF288C098D098C0A03198D0A08008313B6
:1001600003138312640008008316 86010D30850096
:100170008312061083160610831286108316861 0CB
:100180008312061183160611831286118316 8611B7
:100190008312061283160612831286128316 8612A3
:1001A0008312061383160613831286138316 86138F
:1001B0008312051683160512831 27F30A5007F3047
:1001C000A6007F30A7007F30A8007F30A9007F30D5
:1001D000AA007F30AB007F30AC000530A2000130B8
:1001E000A00004309F000730A300A10 11420031CCD
:1001F0004629FF3C031DF62814 20031C4629AD00A8
:100200001420031C4629A4000530A2000230A000DF
:100210004309F002D08472024 08472064002D0843
:1002200003C031D15292408A50064002D08013C8D
:1002300031D1C292408A60064002D08023C031D90
:1002400023292408A70064002D08033C031D2A2944
:100250002408A80064002D08043C031D312924084B
:10026000A90064002D08053C031D38292408AA00B4
:1002700064002D08063C031D3F292408AB006400E0
:100280002D08073C031D46292408AC0025088C00D6
:100290008D010630840001300120 26088C008D017C
```

:1002A00006308400023001202708 8C008D010630C2
:1002B000840004300120280 88C008D010630840061
:1002C000083001202 9088C008D0106308400103090
:1002D00001202A088C008D0106308400 2030012086
:1002E0002B088C008D010630840040300120 2C0842
:0E02F0008C008D0106308400 80300120ED2846
:02400E00F53F7C
:00000001FF

Start the serial servo controller software that was developed in Chapter 6 and experiment with the different facial emotions that can be expressed by positioning the mouth, eyes, and eyebrows. Some of the expressions that I came up with are pictured in **Figure 8.14**.

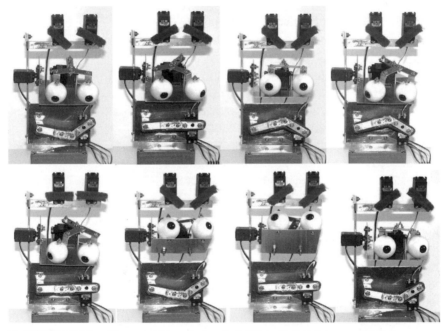

■ **FIGURE 8.14**   *Humanoid head showing different facial expressions.*

Face Mnemonics

As you can see in **Figure 8.14**, the robot can express emotions such as happiness, sadness, disgust, confusion, anger, being bored, etc., with only a minimum set of facial variables. The idea of a machine showing emotion is for the benefit of the humans that will be interacting with the machine. Chernoff first proposed the use of drawing faces with varying features to represent information in 1971. Human

158

beings are trained from birth to attend to the minute details of human faces and to read their expressions. This particular scheme of communication with people seems to be very effective because humans are experienced at recognizing and examining human faces. Even young children can distinguish among large sets of faces after short viewing times. People grow up studying and reacting to faces all the time. Small and barely measurable differences are easily detected and evoke emotional reactions from a long list stored in memory. Relatively large differences go unnoticed in situations where they are not important. The flexibility to disregard noninformative data and to search for useful information is valuable. The ability to relate faces to emotional reactions carries a mnemonic advantage.

Taking It Further

To evolve the head further, you could write a program for the PIC 16F84 microcontroller on the serial servo controller circuit board that would allow it to receive a single serial command that would then produce the proper facial expression. For example, you could create your own commands such as happy, sad, angry, confused, disgusted, that are sent serially from a robot's main controller to the PIC.

Another fun project would be to add a voice to the robot face. A company called RC Systems (www.rcsys.com) produces the V8600A voice synthesizer module. This module converts any ASCII text into speech automatically and requires only a single +5-V supply and a speaker. The module, pictured in **Figure 8.15**, also contains a dual sinusoidal generator, a three-voice musical tone generator, a DTMF (touch-tone) dialer, and the ability to play back sound files. You could coordinate

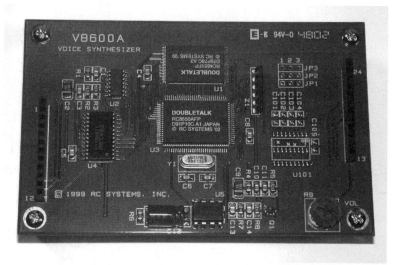

■ **FIGURE 8.15** *V8600A voice synthesizer module by RC Systems, Inc.*

the robot's mouth movements with the speech that is being generated. Use your Visual Basic experience gained in Chapter 6 to develop an application that brings all of these elements together into one complete user interface.

Summary

The humanoid face that was built in this chapter can be used as a method of conveying human emotions through facial expression. This kind of machine-to-human interface serves the purpose of making the experience of interacting with machines much more natural and enjoyable.

160

9

Build Your Own Bipedal Walking Robot

The ability to walk upright on two legs is one of the requirements of a humanoid robot that was determined in Chapter 4. To build a full-scale bipedal humanoid robot, the control system and mechanics would be very complex and expensive. In this chapter we will build a 9-inch tall, autonomous, bipedal walking robot with a PIC 16F819 microcontroller for a brain.

Building the biped robot presented in **Figure 9.1** will allow you to experience the excitement of creating your own artificial lifeform that can walk on two legs, explore its environment, and react to sensory feedback. This bipedal walking robot is unique because it uses two direct current (DC) gear motors contained in one unit to power its legs. One gear motor drives the leg mechanism on the left side of the robot's body and the other gear motor drives the leg mechanism on the right side. Each leg bends at the knee and has a positional sensor (potentiometer) so that walking routines can be accurately programmed. This robot gains stability statically by using a large foot in the form of three sides of a square. It lifts one foot up so that the lateral rods of this foot pass over those of the other, by bending at the knee. The center of gravity of the body is always over the foot that is on the ground. This simplifies programming of the robot's walking gaits because most of the stabilization involved in walking has been taken care of mechanically. The placement of each foot opposite the one that is moving is still very important for this robot to achieve stability while walking.

The bipedal robot controller circuit is designed around the PIC 16F819, which contains 16 input/output (I/O) pins, including five 12-bit analog-to-digital converters. Another feature of this device is a software selectable internal oscillator that can be configured to run between 2 and 8 MHz. With the sophistication of the new programmable interface controller (PIC) microcontrollers, the robot controller board uses fewer components than would have been required in the past.

■ **FIGURE 9.1** *Autonomous bipedal walking robot.*

Mechanical Construction of the Bipedal Walker

The construction of the bipedal walker will begin with the mechanical construction of the body, legs, feet, and the top cover. The parts needed for the mechanical construction are listed in **Table 9.1**.

The body and top cover are constructed using 1/16-inch thick flat aluminum. The construction of the bipedal robot will start with the assembly of the Tamiya twin-motor gearbox. It is available from most hobby robotics suppliers. One supplier is HVW Tech, you can purchase the gearbox directly from its Web site at www.hvwtech.com. The gearbox is sold as a kit and must be assembled before it can be used. **Figure 9.2** shows the entire kit before assembly.

TABLE 9.1 *Parts List for the Bipedal Walker's Mechanical Construction*

Parts	Quantity
1/16-inch thick aluminum stock	8 × 10-foot piece
1/4 × 1/4-inch aluminum stock	6 inches
3/4 × 3/4-inch aluminum square tubing	13 inches
1/2 × 1/8-inch aluminum stock	24 inches
1/2 × 1/2-inch angle aluminum	5 inches
1/4-inch diameter round aluminum tubing	1 inch
6/32 × 1/2-inch machine screws	8
6/32 × 3/4-inch machine screws	6
6/32 × 1-inch machine screws	6
6/32 × 1 1/4-inch machine screws	2
6/32 locking nuts	22
6/32 nylon washers	38
Tamiya twin-motor gear box	1
Heat shrink tubing	2 inches
2-connector female header—2.5 mm spacing	3
3-connector female header—2.5 mm spacing	3

■ **FIGURE 9.2** *Tamiya twin motor gearbox before assembly.*

Assembling the Twin-Motor Gearbox

Take all of the parts out of the box and unfold the instruction sheet. The gearbox has two possible configuration options of standard speed with a gear ratio of 58:1, or low speed with a gear ratio of 203:1. The gearbox will be assembled for use with the low speed option with a gear ratio of 203:1. First, when assembling the gearbox, position a gear hub on each of the two hexagonal output shafts, as shown in **Figure 9.3**. Thread a grub screw into each of the gear hubs with the hex wrench that was supplied with the kit. Use piece M3 to set the proper position of the gear hubs, and then tighten them in place with the hex wrench.

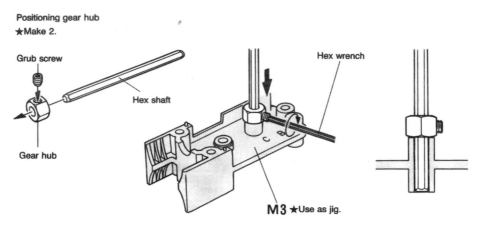

■ **FIGURE 9.3**   *Procedure to attach gear hubs to the hexagonal output shafts.*

Break apart each of the gearbox body sections and plastic spacers from the injection-molded piece and trim off any rough edges with a small knife. Locate the gears, eyelets, screws, and output shafts, then assemble according to **Figure 9.4**.

Place a pinion gear onto the end of each motor shaft so that the end of the shaft is level with the end of the gear. Install each motor in the gearbox by sliding it into place, as shown in **Figure 9.5**. The plastic clips on the gearbox body will snap into place and secure the motors in position.

When the gearbox is complete, mark each shaft at 7/8 of an inch from the body and cut with a hacksaw. Cut two pieces of 2-strand connector wire to a length of 8 inches each. On one end of each of the cables, solder a 2-connector female header and use heat shrink tubing to protect the solder joints. Solder the first wire of the first connector cable to one of the power terminals of the left motor. Solder the second wire of the first connector cable to the remaining power terminal of the left motor. Repeat this same procedure for the second connector cable and the right motor. The finished gearbox, ready for use with the biped robot, is shown in **Figure 9.6**.

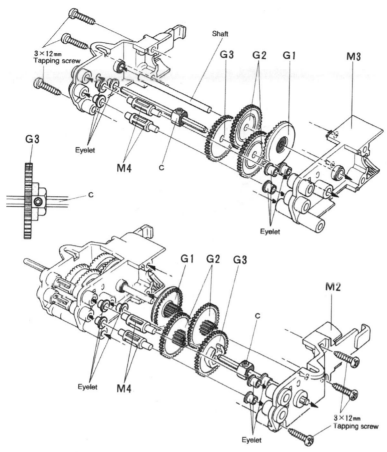

■ **FIGURE 9.4**   *Gearbox assembly diagram.*

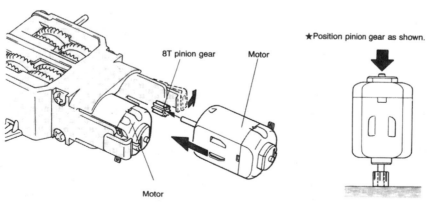

■ **FIGURE 9.5**   *Installing motors in the gearbox.*

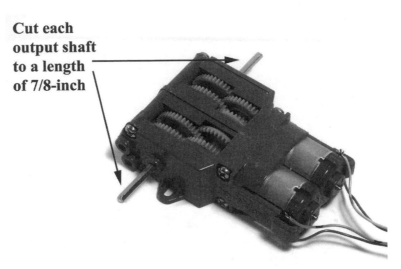

Cut each output shaft to a length of 7/8-inch

■ **FIGURE 9.6**   *Completed twin-motor gearbox with a 203:1 gear reduction.*

Constructing the Chassis

The first step in creating the robot is to construct the aluminum base to which the legs, electronics, gear motors, batteries, and controller circuit board will be fastened. The main body chassis is constructed using a piece of 1/16-inch thick flat aluminum, and is labeled as part A. This will require the use of a hacksaw (or a band saw with a metal cutting blade), a power drill, table vice, and a metal file. Cut and drill a piece of 1/16-inch thick flat aluminum (5 inches wide × 5 inches long) to the dimensions shown in **Figure 9.7**. Drill all of the holes indicated in **Figure 9.7** using a 5/32-inch drill bit, except for the two holes that are marked as being drilled with a 5/16-inch bit. Use a metal file to smooth the edges and remove any burs from the drill holes. Bend the aluminum inward on 90-degree angles, according to the dotted bending lines shown in **Figure 9.7**. Use a bench vise or the edge of a table to bend the pieces.

Locate the two 3-V battery packs (2 × AA) and fasten them to the bottom of the body chassis with two 2/56 × 1/4-inch machine screws and nuts. Use **Figure 9.8** and **Figure 9.9** as a guide when attaching the battery packs to the body chassis. Next, attach the battery clips to the battery packs and wire them together in series to achieve a 6-V DC output by following the wiring guide shown in **Figure 9.8**. Note that the 6-volt output wires are fed through the hole in the bottom of the chassis up to the top side, as indicated in **Figure 9.8**. Use hot glue to fasten the loose wires to the underside of the chassis so that the robot does not trip on them while walking. If the lengths of the wires, measured from the top of the robot chassis, are not at least 4 inches long, then add extension wire. Solder a 2-connector

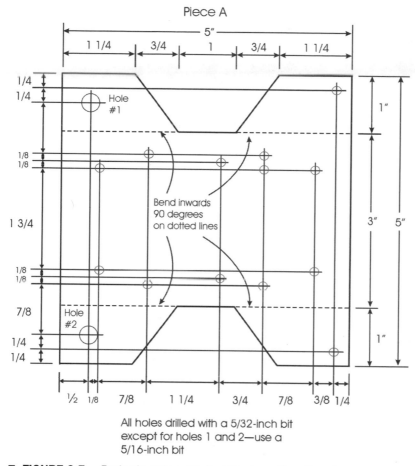

Piece A

All holes drilled with a 5/32-inch bit
except for holes 1 and 2—use a
5/16-inch bit

■ **FIGURE 9.7** *Body chassis cutting, drilling, and bending diagram.*

female header to the end of the 6-V DC output wires. Insulate each of the connections with a 1/2-inch piece of heat shrink tubing.

At this point in the construction, the body chassis with the two 3-V (2 × AA) battery packs fastened to the underside should look like the one shown in **Figure 9.9**.

Locate the two 5 KΩ potentiometers and attach a 5-inch long, 3-strand connector wire to each one. Solder a 3-connector female header to the other end of each wire according to the wiring diagram shown in **Figure 9.10**. Note that the middle terminal of the potentiometer must be connected to the outer terminal of the header. Insulate each of the connections with a 1/2-inch piece of heat shrink tubing. The finished potentiometers are shown in **Figure 9.11**.

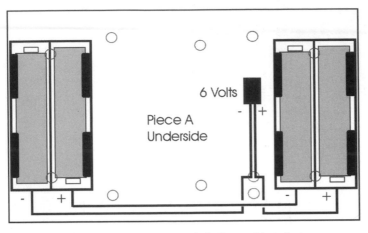

6 Volts

Piece A
Underside

- +

- +

- +

■ **FIGURE 9.8** *Motor power supply battery wiring diagram.*

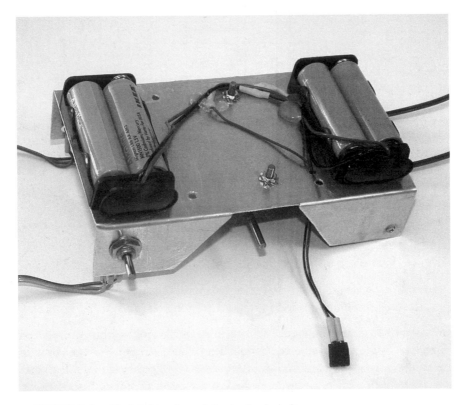

■ **FIGURE 9.9** *Underside view of the body chassis.*

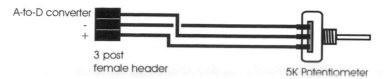

A-to-D converter

−
+

3 post
female header

5K Potentiometer

■ **FIGURE 9.10** *Potentiometer wiring diagram.*

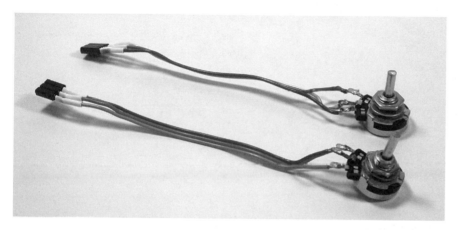

■ **FIGURE 9.11** *Connector wires attached to potentiometers.*

Before the legs are attached to the chassis, each of the potentiometer shafts must be set to their middle positions. This is accomplished by the procedure that follows.

Use **Figure 9.12** as a guide to attach a 5-V DC supply to the outer terminals of the first potentiometer. Attach the leads of a multimeter to the middle terminal and ground so that the voltage can be read. Turn the potentiometer shaft until you get a reading of 2.5 V. Calibrate the second potentiometer using the same procedure.

Now that the gear motors and potentiometers are wired, it is time to attach them to the robot's body chassis. Position the gear motor as shown in **Figure 9.13** and secure it to the chassis using the two machine screws and nuts that came with the motor kit. Mount each of the potentiometers in the 5/16-inch holes at the front of the robot chassis, as shown in **Figure 9.13**. Make sure that the nuts are secured tightly so that the potentiometers do not move out of position when the robot is in operation. Now that the chassis is complete, we will construct the legs, feet, and shaft mounts.

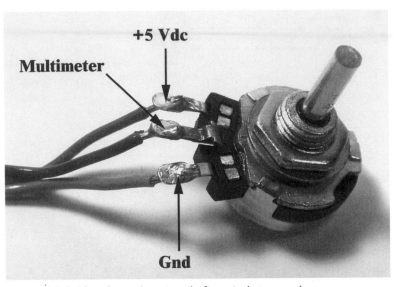

■ **FIGURE 9.12** *Potentiometer shaft centering procedure.*

■ **FIGURE 9.13** *Twin-motor gearbox and potentiometers attached to the chassis.*

Constructing the Legs, Feet, and Shaft Mounts

Cut two pieces of 3/4 × 3/4-inch square aluminum tubing to a size of 4-1/2 inches (pieces A and B), and two pieces to a size of 2 inches (pieces C and D). Use Figure 9.14 to cut and drill the upper and lower leg pieces to the correct dimensions.

Using the 1/2 × 1/8-inch aluminum stock, cut and drill four linkage pieces (E, F, I, and J) according to the dimensions shown in Figure 9.15. Use a 5/32-inch bit to drill the holes. Cut and drill two linkage pieces (G and H) and two ankle pieces (K and L) out of 1/16-inch thick flat aluminum stock according to the dimensions shown in Figure 9.15. Note that the corners should be rounded on pieces G, H, I, and J using a metal file. Use Figure 9.16 as a guide to cut and drill the two foot pieces (M and N) out of 1/2 × 1/2-inch angle aluminum. Use 1/16-inch thick flat aluminum stock to cut, drill, and bend foot pieces (O and P) according to the dimensions shown in Figure 9.16.

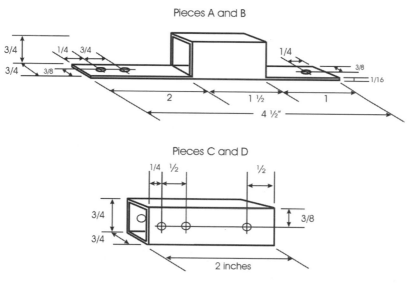

All holes are drilled with a 5/32-inch bit

■ **FIGURE 9.14** *Cutting and drilling dimensions for leg pieces.*

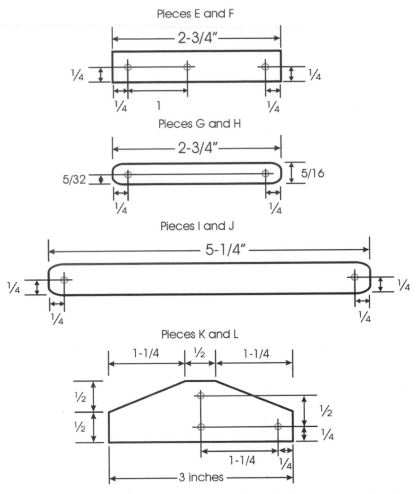

Pieces E and F

2-3/4"

1/4   1/4   1/4   1   1/4

Pieces G and H

2-3/4"

5/32   5/16   1/4   1/4

Pieces I and J

5-1/4"

1/4   1/4   1/4   1/4

Pieces K and L

1-1/4   1/2   1-1/4

1/2   1/2   1/2   1/4

1-1/4   1/4

3 inches

■ **FIGURE 9.15**   *Cutting and drilling dimensions for linkage and ankle pieces.*

Fabricate the motor output shaft mounts and potentiometer shaft mounts using 1/4 × 1/4-inch aluminum square stock according to the dimensions shown in **Figure 9.17**. The motor shaft mounts are labeled as parts Q and R, and the potentiometer shaft mounts are labeled as parts S and T. When the pieces are finished, thread a 6/32 diameter × 3/16-inch set screw in each of the holes that were threaded with the 6/32-inch tap.

Cut two spacers (U and V) to a length of 1/2-inch each using 1/4-inch diameter aluminum tubing, as shown in **Figure 9.17**.

Pieces M and N

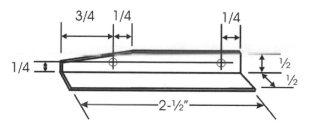

3/4 1/4   1/4

1/4

½
½

2-½"

Pieces O and P

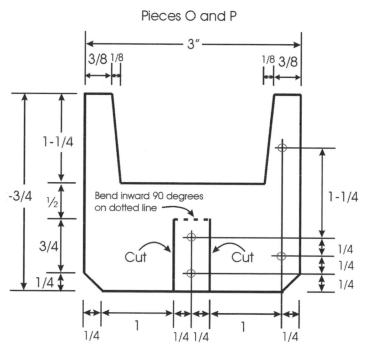

3"

3/8 1/8   1/8 3/8

1-1/4

-3/4   ½

Bend inward 90 degrees
on dotted line

1-1/4

3/4   Cut   Cut

1/4
1/4

1/4   1/4

1   1/4 1/4   1   1/4

■ **FIGURE 9.16** *Cutting, drilling, and bending guide for foot pieces.*

Motor shaft mounts Q and R
2 pieces

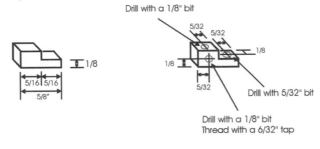

Drill with a 1/8" bit

5/32   5/32

1/8

1/8

5/16 5/16

5/8"

1/8

5/32

Drill with 5/32" bit

Drill with a 1/8" bit
Thread with a 6/32" tap

Potentiometer shaft mounts S and T
2 pieces

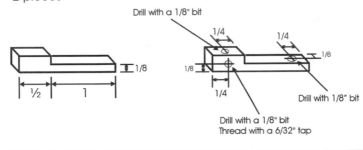

Drill with a 1/8" bit

1/4   1/4

1/8

1/8

1/8

½   1

1/4

Drill with 1/8" bit

Drill with a 1/8" bit
Thread with a 6/32" tap

1/4"

½"

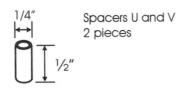

Spacers U and V
2 pieces

■ **FIGURE 9.17** *Motor and potentiometer shaft mounts fabrication diagrams.*

Assembling the Legs

We will begin by assembling the entire left leg, starting with the foot. During the assembly of the leg, refer to **Figures 9.18** and **9.19** for a visual representation. Attach piece M to piece O using two 6/32 × 1/2-inch machine screws and locking nuts. Attach piece K to piece O and piece C using two 6/32 × 1-inch machine screws and locking nuts. Connect the 1-inch end of piece A to piece C using a 6/32 × 1 1/4-inch machine screw and locking nut, with 6/32 nylon washers between each piece. Tighten with enough force to allow the pieces to move freely, but still hold them firmly in place. Attach one end of piece I to piece K using a 6/32 × 1/2-inch machine screw and locking nut, with 6/32 nylon washers between each piece. Tighten with enough force to allow the pieces to move freely, but still hold them firmly in place. Attach piece E to piece A using a 6/32 × 1/2-inch machine screw

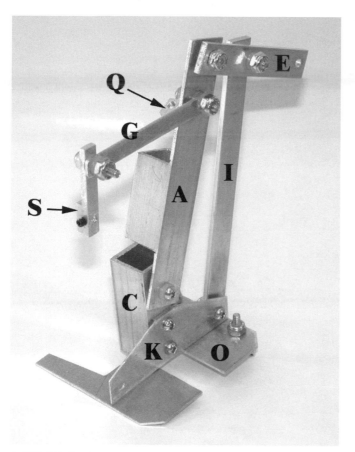

175

■ **FIGURE 9.18** *Left leg assembly diagram—front view.*

and locking nut, and piece E to piece I using a 6/32 × 3/4-inch machine screw and locking nut. Place 6/32 nylon washers between each piece. Tighten with enough force to allow the pieces to move freely, but still hold them firmly in place. Attach piece Q to piece A and piece G with a 6/32 × 3/4-inch machine screw and locking nut, with 6/32 nylon washers between each piece. Tighten with enough force to allow the pieces to move freely, but still hold them firmly in place. Attach piece S to piece G with a 6/32 × 3/4-inch machine screw and locking nut, with 6/32 nylon washers between each piece. Tighten with enough force to allow the pieces to move freely, but still hold them firmly in place.

To assemble the right leg, follow the procedure as stated above with each corresponding piece. Refer to **Figure 9.20** as a guide when assembling the right leg in relation to the left leg.

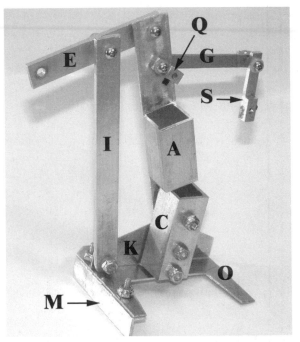

■ **FIGURE 9.19** *Left leg assembly diagram—rear view.*

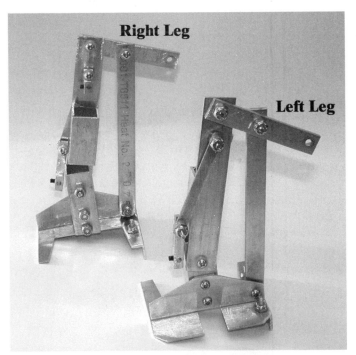

■ **FIGURE 9.20** *Completed right and left legs.*

Attaching the Legs to the Chassis

Refer to **Figure 9.21** while following this procedure. Starting with the robot's left leg, slide motor shaft-mount piece Q onto the motor shaft, and at the same time, slide potentiometer shaft-mount piece S onto the potentiometer shaft, with piece S positioned upward. The end of each shaft should be flush with the outer edge of each mount. Tighten the set screw on the motor shaft mount so that it is secured firmly to the shaft. Place spacer piece U between piece E and the chassis, and then secure in place with a 6/32 × 1-inch machine screw and locking nut, with a nylon washer between piece E and the chassis. Tighten with enough force to allow the pieces to move freely, but still hold them firmly in place. Rotate the motor shaft by hand so that the leg is in its downward and straight position, and then tighten the set screw on the potentiometer shaft mount. Use **Figure 9.21** and the above procedure to attach the right leg to the robot.

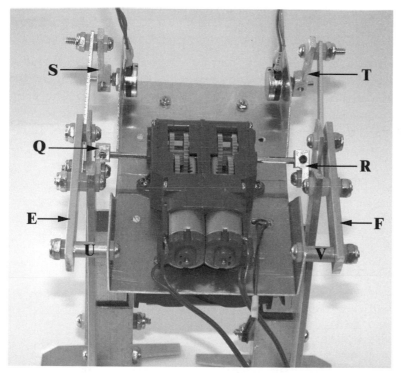

■ **FIGURE 9.21**   *Placement diagram for attaching the legs to the chassis.*

The Main Controller Board

The biped walking robot controller board schematic is shown in **Figure 9.22**. All of the parts needed to create the board are listed in **Table 9.2**. The circuit is designed around Microchip's PIC 16F819 microcontroller. The main part of the circuit is made up of two H-bridge motor controller configurations that consist of two 2N4401 (NPN) and two 2N4403 (PNP) transistors each. The 1N4148 diodes create a voltage path to ground to protect the transistors from any transient high-voltage spikes produced by the DC motors when they are first turned on. The H-bridges are used to control the two DC motors contained in the Tamiya gearbox that drives the legs. The left motor drives the leg mechanism on the left side of the robot's body, and the right motor drives the legs on the right side of the body. By coordinating the movement of each of the legs, the robot is capable of walking forward, walking in reverse, turning to the left, and turning to the right. The regular I/O on Port B pins 0, 1, 2, and 3 are used to control the H-bridge circuits that drive the DC gear motors. Port B pins 4 and 5 are used to control light-emitting diodes. Port B pin 6 is used to output sound to a piezoelectric element. All of the other unused pins have header connectors attached so that they can be used to interface other sensors or output devices that you may want to add during experimentation. Three of the analog-to-digital converters on Port A (pins 0, 1, and 2) of the 16F819 are used to read the voltages produced by the two leg position potentiometers (R2 and R3) and the output voltage produced by the Sharp GP2D12 infrared ranger module. The Sharp GP2D12 ranger is an inexpensive sensor that takes a continuous distance reading, and reports the distance as an analog voltage (0 V to 3 V), with a range of 10 cm

178

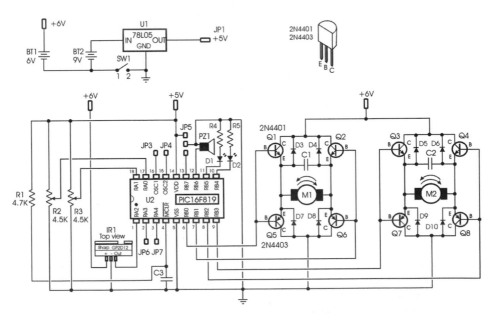

■ **FIGURE 9.22** *The biped robot's main controller board schematic.*

(~4 inches) to 80 cm (~30 inches). The interface is 3-wire with power, ground, and the output voltage. The module requires a JST 3-pin connector, which is included with each detector package. The GP2D12 is shown in **Figure 9.23**. This circuit relies on the PIC 16F819 microcontroller shown in **Figure 9.24,** which functions according to its internal software. Programming the microcontroller will be described after the circuit board is completed and the robot is wired.

Table 9.2   *Main Controller Board Parts List*

Part	Description	Quantity
Semiconductors		
U1	78L05 5-V regulator	1
U2	PIC 16F819 microcontroller	1
Q1–Q4	2N4401 NPN general purpose	4
Q5–Q8	2N4403 PNP general purpose	4
D1, D2	Light-emitting diodes	2
D3–D10	1N4148 diodes	8
Resistors		
R1	4.7 KΩ	1
R2, R3	4.5 KΩ potentiometer with a 1/8-inch diameter and 5/8-inch length shaft	2
R4, R5	470 KΩ	2
Capacitors		
C1, C2, C3, C4, C5	0.1 µfd	5
Miscellaneous		2
Sharp GP2D12 module	Sharp infrared sensor module	1
PZ1	Piezo speaker	1
SW1	On/off toggle switch SPST	1
BT1	3-V battery holder (2 x AA)	2
M1, M2	Tamiya dual-motor gearbox—item 70097	1
Connectors	2-connector female header—2.5 mm spacing	3
JP1-JP10	2-post male header connector—2.5 mm spacing	10
Connectors	3-connector female header—2.5 mm spacing	3
JP11-JP14	3-post male header connector—2.5 mm spacing	4
Ribbon wire	3-strand	8 inches
Ribbon wire	2-strand	16 inches
Hookup wire	18-gauge plastic coated	2 feet
Standoffs	2/56 × 1 1/4-inch	4
Standoffs	2/56 × 1-inch	2
Machine screws for standoffs	2/56 × 1/4-inch	10
BT2	9-V battery connector—PCB mount	1
Heat shrink tubing	1/8-inch diameter	10 inches
Printed circuit board	Available at www.thinkbotics.com	1
Controller board battery	9-V	1
Motor supply batteries	1.5-V type AA	4

179

■ **FIGURE 9.23** *The Sharp GP2D12 infrared range sensor.*

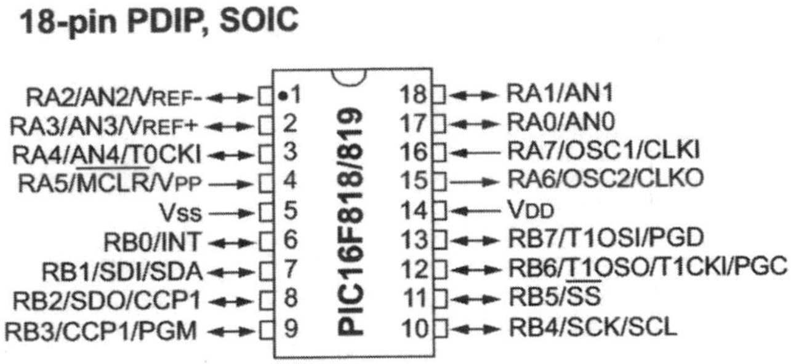

18-pin PDIP, SOIC

RA2/AN2/VREF– ◄►☐ •1		18☐◄► RA1/AN1
RA3/AN3/VREF+ ◄►☐ 2		17☐◄► RA0/AN0
RA4/AN4/T0CKI ◄►☐ 3		16☐◄► RA7/OSC1/CLKI
RA5/MCLR/VPP ──►☐ 4	PIC16F818/819	15☐◄► RA6/OSC2/CLKO
Vss ──►☐ 5		14☐◄── VDD
RB0/INT ◄►☐ 6		13☐◄► RB7/T1OSI/PGD
RB1/SDI/SDA ◄►☐ 7		12☐◄► RB6/T1OSO/T1CKI/PGC
RB2/SDO/CCP1 ◄►☐ 8		11☐◄► RB5/SS
RB3/CCP1/PGM ◄►☐ 9		10☐◄► RB4/SCK/SCL

■ **FIGURE 9.24** *Microchips' PIC 16F819 microcontroller pin diagram.*

Fabricating the Main Controller Printed Circuit Board

The circuit is easiest built by fabricating a circuit board using the artwork shown in **Figure 9.25**. The circuit board can be produced using any method with which you are comfortable. To use the artwork for the photofabricaton process (recommended), photocopy it onto a transparency and follow the directions in Chapter 2. When you are ready to expose the copper board, orient the transparency exactly as shown in **Figure 9.25**. The exact size of the board should be 5-1/16 inches × 2-1/2 inches. If you are going use the iron-on transfer method, you will need to scan the foil pattern, and then mirror the image so that the artwork is properly oriented when it is printed onto the transfer sheet and then ironed onto the copper board.

The finished printed circuit board is also available to purchase from the author's Web site, www.thinkbotics.com. You can also download the image file free at the same location. If you don't want to fabricate a printed circuit board, the circuit is simple enough to construct on a 5-1/16 × 2-1/2 inch piece of standard perforated

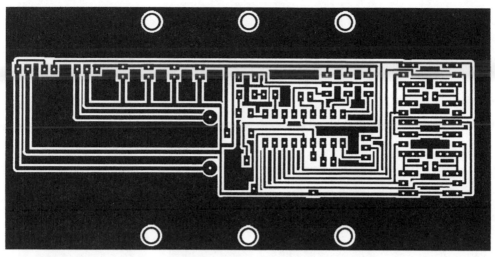

■ **FIGURE 9.25**   *Printed circuit board artwork for the main controller board.*

circuit board, using point-to-point wiring, although the details of that procedure will not be covered.

Once the circuit board has been etched, drilled, and cut, use **Figure 9.26** and **Table 9.2** to place the parts on the board. Solder an 18-pin IC socket where part U2 (PIC 16F819) is shown. The PIC will need to be programmed before it is inserted into the socket (more about programming later). Solder all parts in place after they have been positioned on the board. Attach four 2/56 × 1 1/4-inch standoffs to the outer mounting holes on the copper side of the circuit board, and two 2/56 × 1-inch standoffs on the top of the board to the middle holes, as shown in **Figure 9.27**. Mount the board to the back of the robot with the 9-V battery holder situated above the dual motor gearbox, as shown in **Figure 9.28**. Use **Figure 9.29** as a guide

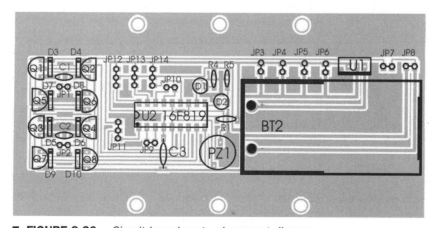

■ **FIGURE 9.26**   *Circuit board parts placement diagram.*

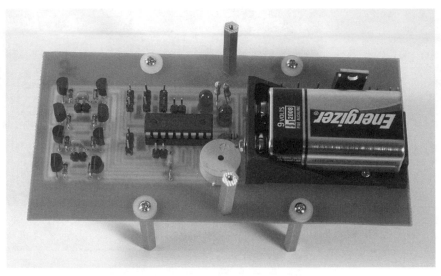

■ **FIGURE 9.27** *Finished controller board.*

■ **FIGURE 9.28** *Controller board mounted to the chassis.*

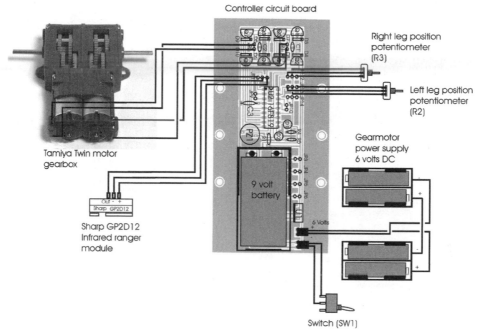

Controller circuit board

Right leg position
potentiometer
(R3)

Left leg position
potentiometer
(R2)

Gearmotor
power supply
6 volts DC

9 volt
battery

6 Volts

Tamiya Twin motor
gearbox

Sharp GP2D12

Sharp GP2D12
Infrared ranger
module

Switch (SW1)

■ **FIGURE 9.29**   *Robot electronics wiring connection diagram.*

to connect all of the components to the controller circuit board. The power switch (SW1) will eventually be mounted in the 1/4-inch hole on the robot's top cover, which will be added when the programming is complete. The GP2D12 will also be attached to the top cover a little later, so attach it to JP11 and tape it to the front of the robot during the testing and programming phase. Note that the robot's left leg potentiometer (R2) is attached to JP13, and the right potentiometer (R3) is attached to JP12. The Sharp GP2D12 is attached to JP11. The left motor is connected to JP2, and the right motor is connected to JP1.

Programming the PIC 16F819 Microcontroller

To program the microcontroller, you will need a hardware programmer such as the EPIC Plus programmer and the PicBasic Pro compiler that were described in Chapter 3. Both the compiler and programmer are available from microEngineering Labs (www.melabs.com). The program listings shown are produced for use with the PicBasic Pro compiler, but could be translated to work with any PIC microcontroller compiler. When the code has been compiled, a standard 8-bit Merged Intel HEX (.hex) file is created that can be used with any PICmicro programmer. This machine code file is then loaded into the EPIC Plus programming software and used to program the PIC.

When the PIC 16F819 has been programmed and inserted into the 18-pin socket on the controller board, it will start executing the code when power is supplied. **Program 9.1** is called test-biped.bas, and it will be used to test the robot's functions. Once the program has been compiled, program the PIC 16F819 with the test-biped.hex file listed in **Program 9.2**. For your convenience you can download the Basic and Hex files for this project from www.thinkbotics.com. When the 16F819 is programmed, insert it into the 18-pin socket on the controller board, with pin 1 facing the notch in the socket located closest to the transistors of the H-bridge section of the circuit (see **Figure 9.26**). Make sure that a fresh 9-V battery and four 1.5-V AA batteries are placed in the battery holders. When the power is turned on, the robot should produce sound from the piezo element, run each leg forward, and then reverse while flashing the light-emitting diodes (LEDs) on and off in sequence. When it has finished rotating the legs and flashing, the LED's program execution will go into a loop to test the infrared ranger.

To test the ranger, move your hand in front of the robot at a distance of 4 to 5 inches. The microcontroller will output sound to the piezo element when your hand is within the specified distance. Because the output of this sensor module is nonlinear, there is a dead zone of 2 inches directly in front of the robot. This is not a problem because the robot walks at a relatively slow speed and the program is looking at a wide range of values. If you want to accurately interpret the nonlinear voltages produced by the sensor, you can write a routine that uses a lookup table to correlate all of the voltages to actual distances. Note that there is a subroutine included at the end of the program called *Calibrate*. You can use this routine to read the actual values from the analog-to-digital converters to which the potentiometers and the infrared sensor module are connected. If you find that the left and right legs are not alternating their movement, then put the *Calibrate* subroutine into a loop at the start of the program. Attach a serial LCD module to Port B pin 7, and adjust each potentiometer so that a value of 128 is displayed when each leg is in its downward and straight position.

■ **PROGRAM 9.1** *test-biped.bas program listing.*

```
'_____

' Name    : Test-biped.bas
' Compiler : PicBasic Pro MicroEngineering Labs
' Notes    : biped Robot test program
'_____


@ DEVICE PIC16F819, INTRC_OSC_NOCLKOUT, WDT_OFF, LVP_OFF, PWRT_ON,
    PROTECT_OFF, BOD_OFF

include "modedefs.bas"
```

```
TRISA = %00011111
TRISB = %00000000
DEFINE OSC 8
OSCCON = $70


M1            Var PORTB.0
M2            Var PORTB.1
M3            Var PORTB.2
M4            Var PORTB.3
LED1          VAR PORTB.4
LED2          var PORTB.5
PIEZO         var PORTB.6
LCD           VAR PORTB.7
LCD_BAUD      CON N2400

IOW M1
LOW M2
LOW M3
LOW M4
low LED1
low LED2
low PIEZO


LEFT_POT      VAR BYTE
RIGHT_POT     VAR BYTE
LEG_STOP_F    VAR BYTE
LEG_STOP_R    VAR BYTE
INFRARED      VAR BYTE
VARIANCE      VAR BYTE


LEG_STOP_F = 128
LEG_STOP_R = 119
VARIANCE = 1

' Set up the analog to digital converters

DEFINE ADC_BITS 8          ' Set number of bits in result
DEFINE ADC_CLOCK 1         ' Set clock source (1 = internal 8 MHz, 3 = rc)
DEFINE ADC_SAMPLEUS 50     ' Set sampling time in microseconds
ADCON1 = 0                 ' Set PortA pins to analog
sound PIEZO,[100,10,90,5,80,5,110,10]
```

```
START:

   sound PIEZO,[100,5,110,5]
   LOW  LED1
   HIGH LED2
   GOSUB LEFT_FORWARD

   sound PIEZO,[80,5,90,5]
   LOW  LED2
   HIGH LED1
   GOSUB LEFT_REVERSE

   sound PIEZO,[100,5,110,5]
   LOW  LED1
   HIGH LED2
   GOSUB RIGHT_FORWARD

   sound PIEZO,[80,5,90,5]
   LOW  LED2
   HIGH LED1
   GOSUB RIGHT_REVERSE

SENSOR: ADCIN 2,INFRARED

   IF INFRARED > 100 AND INFRARED < 130 THEN
      sound PIEZO,[100,10,90,5,100,5,110,10,80,20,90,20]
   ENDIF

GOTO SENSOR

END

' motor control subroutines start here
'_____

RIGHT_FORWARD:

   RIGHT_POT = 0
   lOW M1
   LOW M2

   HIGH M2
   PAUSE 400
```

```
    WHILE RIGHT_POT < (LEG_STOP_F - VARIANCE) OR RIGHT_POT >
        (LEG_STOP_F + VARIANCE)
      ADCIN 1,RIGHT_POT
    WEND
    LOW M2
RETURN


RIGHT_REVERSE:

  RIGHT_POT = 0
  lOW M1
  LOW M2

  HIGH M1
  PAUSE 300
  WHILE RIGHT_POT < (LEG_STOP_R - VARIANCE) OR RIGHT_POT >
        (LEG_STOP_R + VARIANCE)
      ADCIN 1,RIGHT_POT
    WEND
      LOW M1
RETURN


LEFT_FORWARD:

  LEFT_POT = 0
  lOW M3
  LOW M4

  HIGH M3
  PAUSE 400
  WHILE LEFT_POT < (LEG_STOP_F - VARIANCE) OR LEFT_POT >
        (LEG_STOP_F + VARIANCE)
      ADCIN 0,LEFT_POT
    WEND
      LOW M3
RETURN

LEFT_REVERSE:

  LEFT_POT = 0
  lOW M3
  LOW M4
```

```
        HIGH M4
        PAUSE 300
        WHILE LEFT_POT < (LEG_STOP_R - VARIANCE) OR LEFT_POT >
            (LEG_STOP_R + VARIANCE)
          ADCIN 0,LEFT_POT
        WEND
          LOW M4
RETURN


' LCD display analog-to-digital converter values
'_____


CALIBRATE:

    ADCIN 0,LEFT_POT        ' read A/D converter - porta.pin 0
    serout LCD,LCD_BAUD,[254,128,"L:",#LEFT_POT," "]
    ADCIN 1,RIGHT_POT       ' read A/D converter - porta.pin 1
    serout LCD,LCD_BAUD,[254,134,"R:",#RIGHT_POT," "]
    ADCIN 2,INFRARED        ' read A/D converter - porta.pin 2
    serout LCD,LCD_BAUD,[254,192,"IR:",#INFRARED," "]

RETURN
```

■ **PROGRAM 9.2**   *test-biped.hex file listing.*

```
:100000001529A501A400B7172730A300103014202C
:100010000330A300E8301420A30164301420A301AE
:100020000A30142024081F28A2002508A100240853
:10003000A000F4202008031DB713B71B0800303EB2
:10004000A6003A0884000930A70003102C20A60C53
:10005000A70B262803142C288413B71D3B2800085F
:100060003804371D3806800084170008380403 1C44
:100070003806800046280008380403 1C3806371963
:100080003806800084173809800546283 70D063960
:10009000A0004F20A100A00A4F200000BC28003083
:1000A0008A002008820701348B3403342B34003457
:1000B00052340C34EF34A3003A08840038098B2002
:1000C0008413A30803191029F030A50022088038F2
:1000D000A400F030A5030319A5000319A303031915
:1000E00010292008A6002108A7001630BB202608EA
:1000F000A0002708A1009F20030120183808A21F94
:100100003808A20803190301A40F8828800666286E
:10011000892800006928841780051029A000A00DF7
```

:10012000A00D200D383941389F000030A100323039
:10013000BC201F151F199A28A1011E08102921088B
:1001400020040319A00A8030201AA1062019A10654
:10015000A018A106210DA00DA10D1029A301A20038
:10016000FF30A207031CA307031C10290330A100C2
:10017000E330BC20B028A101F43EA000A109FE306C
:10018000031CC528A0070318C228A0076400A10FFC
:10019000C22800002018CC282018CE280800A10171
:1001A000A301A2000130D928A101A301A2000430BB
:1001B000D928A80023082102031DE02822082002D4
:1001C0000043003180130031902302805031DFF30E5
:1001D00010290038031DFF300405031DFF301029CE
:1001E0000404031DFF301029A501A4011030A6004E
:1001F000210DA40DA50D2208A4022308031C230F22
:100200000A50203180A292208A40723080318230FAC
:10021000A5070310A00DA10DA60BF8282008102992
:10022000831303138312640008008316 1F308500B4
:10023000860170308F008312061083160610831219
:10024000861083168610831206118316061183 12F8
:10025000861183168611831206128316061283 12E4
:10026000861283168612831206138316061383 12D0
:100270008030BE007730BF000130C10083169F017F
:100280006308312BA004030B8006430A2000A3051
:100290005B205A30A20005305B205030A2000530B0
:1002A0005B206E30A2000A305B200630BA0040307E
:1002B000B8006430A20005305B206E30A20005302B
:1002C0005B2006128316061283128616831686128 8
:1002D000831260220630BA004030B8005030A200CD
:1002E0005305B205A30A20005305B208612831651
:1002F0008612831206168316061283 12A022063077
:100300 00BA004030B8006430A20005305B206E3087
:10031000A20005305B20061283160612831286 1691
:10032000831686128312E0210630BA004030B800EE
:100330005030A20005305B205A30A20005305B200F
:10034000861283168612831206168316061283 12ED
:100350000202202308E20BC003C08A0006430CF2058
:10036000B2003C08A0008230D42B40032088400DF
:100370003408E920B400B5006400340835040319DA
:10038000DD290630BA004030B8006430A2000A30DF
:100390005B205A30A20005305B206430A20005309B
:1003A0005B206E30A2000A305B205030A200143077
:1003B0005B205A30A20014305B20A9296300DE299B
:1003C000C00106108316061083128610831686104D

:1003D00083128614831686108312013 0A300903096
:1003E000AF2041083E02B200B301031CB303400832
:1003F000A000A1013308A3003208D620B2003E08B5
:100400004107B400B501B50D4008A000A1013508B1
:10041000A3003408D120B400320884003408F0204E
:10042000B400B5006400340835040319 1B2A0130F8
:100430008E20C000F1298610831686108312080 0D2
:10044000C0010610831606108312861083168610CC
:1004500083120614831606108312013 0A3002C3079
:10046000AF2041083F02B200B301031CB3034008B0
:10047000A000A1013308A3003208D620B2003F0833
:10048000410 7B400B501B50D4008A000A101350831
:10049000A3003408D120B400320884003408F020CE
:1004A000B400B500640034083504031 95B2A013038
:1004B0008E20C000312A0610831606108312080011
:1004C000BD010611831606118312861183168611 4B
:1004D00083120615831606118312013 0A300903093
:1004E000AF2041083E02B200B301031CB3033D0834
:1004F000A000A1013308A3003208D620B2003E08B4
:100500004107B400B501B50D3D08A000A1013508B3
:10051000A3003408D120B400320884003408F0204D
:10052000B400B500640034083504031 99B2A003078
:100530008E20BD00712A0611831606118312080051
:10054000BD010611831606118312861183168611 CA
:1005500083128615831686118312013 0A3002C3076
:10056000AF2041083F02B200B301031CB3033D08B2
:10057000A000A1013308A3003208D620B2003F0832
:100580004107B400B501B50D3D08A000A101350833
:10059000A3003408D120B400320884003408F020CD
:1005A000B400B5006400340835040319 DB2A0030B8
:1005B0008E20BD00B12A8611831686118312080091
:1005C00000030 8E20BD000630BA008030B800043004
:1005D000B700FE302020803020204C3020203A30E0
:1005E00020203D0801202030202020302020013014
:1005F0008E20C0000630BA008030B8000430B7004A
:1006000 0FE3020208630202 0523020203A3020201A
:100610004008012020302020203020200 2308E2071
:10062000BC000630BA008030B8000430B700FE309D
:100630002020C0302020493020205 23020203A3065
:1006400 020203C0801202030202020302020080 0DD
:02400E00303F41
:00000001FF

When the test-biped program is working correctly, compile the biped-explore.bas code listed in **Program 9.3**. Program the PIC 16F819 with the biped-explore.hex file listed in **Program 9.4**. When this program executes, the robot will walk forward while monitoring the values received from the infrared sensor module. If an obstacle is detected, the robot will walk in reverse away from the obstacle, and then turn to the left or to the right. The robot alternates the direction it turns each time an obstacle is sensed.

■ **PROGRAM 9.3** *Biped-explore.bas program listing.*

```
'_____

' Name      : Biped-explore.bas
' Compiler  : PicBasic Pro MicroEngineering Labs
' Notes     : biped Robot exploration program
'_____


@ DEVICE PIC16F819, INTRC_OSC_NOCLKOUT, WDT_OFF, LVP_OFF, PWRT_ON,
    PROTECT_OFF, BOD_OFF

include "modedefs.bas"

TRISA = %00011111
TRISB = %00000000
DEFINE OSC 8
OSCCON = $70


M1          Var PORTB.0
M2          Var PORTB.1
M3          Var PORTB.2
M4          Var PORTB.3
LED1        VAR PORTB.4
LED2        var PORTB.5
PIEZO       var PORTB.6
LCD         VAR PORTB.7
LCD_BAUD    CON N2400

lOW M1
LOW M2
LOW M3
LOW M4
low LED1
low LED2
```

```
low PIEZO

LEFT_POT        VAR BYTE
RIGHT_POT       VAR BYTE
LEG_STOP_F      VAR BYTE
LEG_STOP_R      VAR BYTE
INFRARED        VAR BYTE
VARIANCE        VAR BYTE
TEMP            VAR BYTE
FLAG            VAR bit

LEG_STOP_F = 128
LEG_STOP_R = 119
VARIANCE = 1
FLAG = 0

' Set up the analog to digital converters

DEFINE ADC_BITS 8           ' Set number of bits in result
DEFINE ADC_CLOCK 1          ' Set clock source (1 = internal 8 MHz, 3 = rc)
DEFINE ADC_SAMPLEUS 50      ' Set sampling time in microseconds
ADCON1 = 0                  ' Set porta pins to analog
sound PIEZO,[100,10,90,5,80,5,110,10]

START:

    sound PIEZO,[100,5,110,5]
    LOW  LED1
    HIGH LED2
    GOSUB LEFT_FORWARD

    sound PIEZO,[80,5,90,5]
    LOW  LED2
    HIGH LED1
    GOSUB RIGHT_FORWARD

    ADCIN 2,INFRARED

    IF INFRARED > 100 AND INFRARED < 130 THEN
        sound PIEZO,[100,10,90,5,100,5,110,10,80,20,90,20]
        FLAG = FLAG +1
        FOR TEMP = 1 TO 5
            GOSUB LEFT_REVERSE
```

```
            GOSUB RIGHT_REVERSE
        NEXT TEMP
     IF FLAG Then
        FOR TEMP = 1 TO 5
            GOSUB LEFT_REVERSE
            GOSUB RIGHT_FORWARD
        NEXT TEMP
     Else
        FOR TEMP = 1 TO 5
            GOSUB RIGHT_REVERSE
            GOSUB LEFT_FORWARD
        NEXT TEMP
     Endif
     ENDIF


GOTO START


END


' motor control subroutines start here
'_____
```

```
RIGHT_FORWARD:

  RIGHT_POT = 0
  LOW M1
  LOW M2

  HIGH M2
  PAUSE 400
  WHILE RIGHT_POT < (LEG_STOP_F - VARIANCE) OR RIGHT_POT >
       (LEG_STOP_F + VARIANCE)
    ADCIN 1,RIGHT_POT
  WEND
  LOW M2
RETURN

RIGHT_REVERSE:

  RIGHT_POT = 0
  LOW M1
  LOW M2
```

```
        HIGH M1
        PAUSE 300
        WHILE RIGHT_POT < (LEG_STOP_R - VARIANCE) OR RIGHT_POT >
            (LEG_STOP_R + VARIANCE)
          ADCIN 1,RIGHT_POT
        WEND
          LOW M1
     RETURN

   LEFT_FORWARD:

      LEFT_POT = 0
      LOW M3
      LOW M4

      HIGH M3
      PAUSE 400
      WHILE LEFT_POT < (LEG_STOP_F - VARIANCE) OR LEFT_POT >
          (LEG_STOP_F + VARIANCE)
        ADCIN 0,LEFT_POT
      WEND
        LOW M3
   RETURN

   LEFT_REVERSE:

      LEFT_POT = 0
      LOW M3
      LOW M4

      HIGH M4
      PAUSE 300
      WHILE LEFT_POT < (LEG_STOP_R - VARIANCE) OR LEFT_POT >
          (LEG_STOP_R + VARIANCE)
        ADCIN 0,LEFT_POT
      WEND
        LOW M4
   RETURN

   ' LCD display analog-to-digital converter values
   '_____
```

CALIBRATE:

```
ADCIN 0,LEFT_POT        ' read A/D converter - porta.pin 0
serout LCD,LCD_BAUD,[254,128,"L:",#LEFT_POT," "]
ADCIN 1,RIGHT_POT       ' read A/D converter - porta.pin
serout LCD,LCD_BAUD,[254,134,"R:",#RIGHT_POT," "]
ADCIN 2,INFRARED        ' read A/D converter - porta.pin 2
serout LCD,LCD_BAUD,[254,192,"IR:",#INFRARED," "]
```

RETURN

■ **PROGRAM 9.4**   *Biped-explore.hex file listing.*

```
:100000001529A501A400B7172730A300103014202C
:10001000330A300E8301420A30164301420A301AE
:100020000A30142024081F28A2002508A100240853
:10003000A000F1202008031DB713B71D0800303EB2
:10004000A6003A0884000930A70003102C20A60C53
:10005000A70B262803142C288413B71D3B2800085F
:100060003804371D3806800084170008380403
:10005000A70B262803142C288413B71D3B2800085F
:100060003804371D38068000841700083804031C44
:100070003806800046280008380403 1C3806371963
:100080003806800084173809800546 28370D063960
:10009000A0004F20A100A00A4F200000BC28003083
:1000A0008A002008882070134 8B3403342B34003457
:1000B00052340C34EF34A3003A08840038098B2002
:1000C0008413A30803191029F030A50022088038F2
:1000D000A400F030A5030319A5000319A303031915
:1000E00010292008A6002108A7001630BB202608EA
:1000F000A0002708A1009F20030120183808A21F94
:100100003808A20803190301A40F8828800666286E
:1001100089280000692884178005 1029A000A00DF7
:10012000A00D200D383941389F000030A100323039
:10013000BC201F151F199A28A1011E08102921088B
:10014000020040319A00A8030201AA1062019A10654
:10015000A018A106210DA00DA10D1029A301A20038
:10016000FF30A207031CA307031C10290330A100C2
:10017000E330BC20B028A101F43EA000A109FE306C
:10018000031CC528A0070318C228A0076400A10FFC
:10019000C2280000201 8CC282018CE280800A10171
:1001A000A301A2000130D928A101A301A2000430BB
:1001B000D928A80023082102031DE02822082002D4
:1001C000043003180130031902302805031DFF30E5
:1001D000010290038031DFF300405031DFF301029CE
```

195

:1001E0000404031DFF301029A501A4011030A6004E
:1001F000210DA40DA50D2208A4022308031C230F22
:10020000A50203180A292208A40723080318230FAC
:10021000A5070310A00DA10DA60BF8282008102992
:100220008313031383126400080083161F308500B4
:10023000860170308F008312061083160610831219
:10024000861083168661083120611831606118312F8
:10025000861183168661183120612831606128312E4
:10026000861283168661283120613831606138312D0
:100270008030BE007730BF000130C30040108316CD
:100280009F0106308312BA004030B8006430A200EB
:100290000A305B205A30A20005305B205030A200AB
:1002A00005305B206E30A2000A305B200630BA00B9
:1002B0004030B8006430A20005305B206E30A200F0
:1002C00005305B20061283160612831286168316EB
:1002D0008612831263220630BA004030B8005030D4
:1002E000A20005305B205A30A20005305B20861248
:1002F0008316861283120616831606128312E321D2
:1003000002308E20BC003C08A0006430CF20B20038
:100310003C08A0008230D420B400320884003408A5
:10032000E920B400B5006400340835040319E0295D
:100330000630BA004030B8006430A2000A305B20BA
:100340005A30A20005305B206430A20005305B20EB
:100350006E30A2000A305B205030A20014305B20C7
:100360005A30A20014305B208310401C8314831881
:100370004014831C40100130C20064000630420269
:100380000318C629A3222322C20FBD296400401CE2
:10039000D5290130C2006400063042020318D42976
:1003A000A322E321C20FCB29E0290130C20064005F
:1003B00063042020318E02923226322C20FD72904
:1003C00056296300E129C101061083160610831225
:1003D0008610831686108312861483168610831265
:1003E0000130A3009030AF2043083E02B200B301B9
:1003F00031CB3034108A000A1013308A300320885
:10040000D620B2003E084307B400B501B50D41083F
:10041000A000A1013508A3003408D120B40032089F
:1004200084003408F020B400B500640034083504BA
:1004300003191E2A01308E20C100F429861083166C
:10044000861083120800C10106108316061083125D
:1004500086108316861083120614831606108312E4
:100460000130A3002C30AF2043083F02B200B3019B
:1004700031CB3034108A000A1013308A300320804
:10048000D620B2003F084307B400B501B50D4108BE

:10049000A000A1013508A3003408D120B40032081F
:1004A00084003408F020B400B5006400340835043A
:1004B00003195E2A01308E20C100342A06108316EB
:1004C000061083120800BD010611831606118312 5F
:1004D0008611831686118312061583160611831260
:1004E0000130A3009030AF2043083E02B200B301B8
:1004F000031CB3033D08A000A1013308A300320888
:10050000D620B2003E084307B400B501B50D3D0842
:10051000A000A1013508A3003408D120B40032089E
:1005200084003408F020B400B500640034083504B9
:1005300003199E2A00308E20BD00742A06118316EE
:1005400006118312080 0BD010611831606118312DD
:100550008611831686118312861583168611 8312DF
:100560000130A3002C30AF2043083F02B200B3019A
:10057000031CB3033D08A000A1013308A300320807
:10058000D620B2003F084307B400B501B50D3D08C1
:10059000A000A1013508A3003408D120B40032081F
:1005A00084003408F020B400B50064003408350439
:1005B0000319DE2A00308E20BD00B42A861183166E
:1005C0008611831208000 0308E20BD000630BA006C
:1005D0008030B8000430B700FE302020803020206A
:1005E0004C3020203A3020203D08012020302020AF
:1005F0002030202001308E20C1000630BA0080302B
:100600000B8000430B700FE30202086302020523061
:100610000203A3020204108012020302020203 0A6
:1006200002002308E20BC000630BA008030B80096
:100630000430B700FE302020C03020204930202078
:100640000523020203A3020203C0801202030202049
:0606500020302020 08000C
:02400E00303F41
:00000001FF

Now that the biped can explore its environment and avoid obstacles, a top cover to mount the GP2D12 infrared sensor and power switch will be constructed. Use **Figures 9.30** and **9.31** to construct pieces W, X, and Y out of 1/16-inch thick aluminum. Also, cut and drill four connector pieces labeled Z from 1/16-inch thick aluminum, as shown in **Figure 9.31**. Use rivets and the connector pieces to join piece W to piece X, and piece Y to piece X. Mount the GP2D12 infrared sensor to piece W with the mounting fasteners that came with the sensor. Drill a 1/4-inch hole in the top left side at the back of the cover, and then mount the switch in the 1/4-inch hole. The completed cover is shown in **Figure 9.32**. Mount the cover to the top of the robot and secure in place on the two 1/2-inch standoffs. The finished robot is shown in **Figure 9.33**.

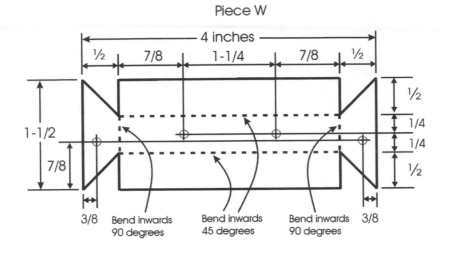

Piece W

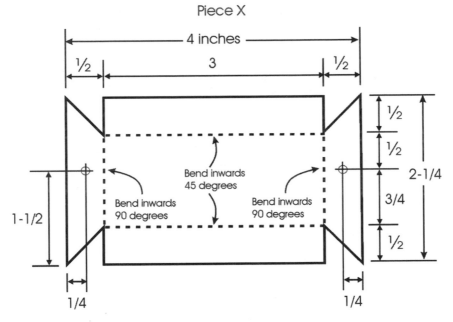

Piece X

■ FIGURE 9.30 *Cutting, drilling, and bending diagram for cover pieces.*

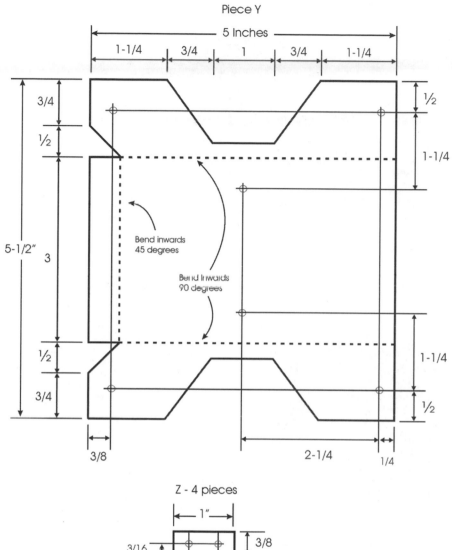

Piece Y

Z - 4 pieces

■ FIGURE 9.31 *Cutting, drilling, and bending diagram for top cover pieces.*

■ **FIGURE 9.32** *Completed top cover with sensor and switch mounted.*

■ **FIGURE 9.33** *Completed bipedal walking robot project.*

Summary

This concludes the construction and programming of the bipedal walking robot. Much more can be done with this robot than what has been covered. Some of the things that can be added to the robot are an ultrasonic range finder, a remote control system, wireless video link, a data link, electronic compass, and light sensors, to name a few. If you do want to add these things and continue experimenting with walking robots, I recommend that you pick up my first two books. These books are titled *Insectronics: Build Your Own Six Legged Walking Robot* (ISBN: 0-07-141241-7) and *Amphibionics: Build Your Own Biologically Inspired Robots* (ISBN: 0-07-141245-x) shown in **Figure 9.34**.

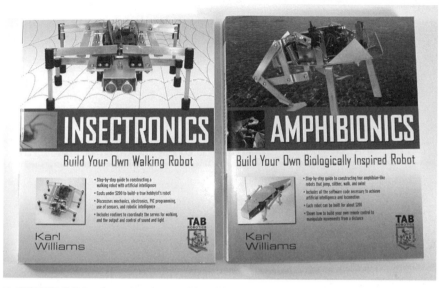

■ **FIGURE 9.34** Insectronics *and* Amphibionics *robotics books.*

Index

204

209

211

About the Author

Karl P. Williams is an independent robotics researcher, electronics experimenter, and software developer. He is with AGFA HealthCare Informatics, a leading medical imaging software company. He is the author of three robotics books titled *Insectronics: Build Your Own Six Legged Walking Robot* (ISBN: 0-07-141241-7), *Amphibionics: Build Your Own Biologically Inspired Robots* (ISBN: 0-07-141245-X), and *Build Your Own Humanoid Robots* (ISBN:0-07-142274-9), all published by McGraw-Hill. A resident of Ontario, Canada, he has written for the magazines *Nuts and Volts*, *SERVO*, and *Conformity*. He hosts a couple of robotics Web sites at (http://home.golden.net/~kpwillia) and (www.thinkbotics.com).